A Life on Our Planet

A Life on Our Planet

My Witness Statement
and A Vision for the Future

David Attenborough

with Jonnie Hughes

GRAND CENTRAL
PUBLISHING

NEW YORK BOSTON

With thanks to WWF for the scientific and conservation work informing this book and the accompanying film.

Copyright © 2020 by David Attenborough Productions Ltd.

Cover copyright © 2020 by Hachette Book Group, Inc.

Grand Central Publishing
Hachette Book Group
1290 Avenue of the Americas, New York, NY 10104
grandcentralpublishing.com
twitter.com/grandcentralpub

Published in Great Britain by Witness Books, an imprint of Ebury Publishing, in 2020.
First American Edition: October 2020

Grand Central Publishing is a division of Hachette Book Group, Inc. The Grand Central Publishing name and logo is a trademark of Hachette Book Group, Inc.

The publisher is not responsible for websites (or their content) that are not owned by the publisher.

The Hachette Speakers Bureau provides a wide range of authors for speaking events. To find out more, go to www.hachettespeakersbureau.com or call (866) 376-6591.

Illustrations and graphics by Lizzie Harper and Meghan Spetch

Print book interior design by seagulls.net

Library of Congress Cataloging-in-Publication Data has been applied for.

ISBNs: 978-1-5387-1998-5 (hardcover), 978-1-5387-2000-4 (ebook)

Printed in the United States of America

LSC-C

10 9 8 7 6 5 4 3 2 1

CONTENTS

Our Greatest Mistake

Pripyat in the Ukraine is a place unlike anywhere else I have been. It is a place of utter despair.

On the face of it, it seems quite a pleasant town, with avenues, hotels, a square, a hospital, parks with fairground rides, a central post office, a railway station. It has several schools and swimming pools, cafés and bars, a restaurant by the river, shops, supermarkets and hairdressers, a theatre and a cinema, a dance hall, gymnasiums and a football stadium with an athletics track. It has all the amenities we humans have brought into existence to give us a content and comfortable life – all the elements of our homemade habitat.

Surrounding the town's cultural and commercial centre, are the apartments. There are 160 towers, built at specified angles to a well-considered grid of roads. Each apartment has its own balcony. Each tower its own laundry. The tallest towers reach almost 20 storeys high, and each is crowned with a giant iron hammer and sickle, the emblem of the town's creators.

Pripyat was built by the Soviet Union, in one busy period of construction in the 1970s. It was the designed, perfect home for almost 50,000 people, a modernist utopia to suit the very best engineers and scientists in the Eastern Bloc, together with their young families. Amateur film footage from the early 1980s shows them, smiling, mingling and pushing prams on the wide boulevards, taking ballet classes, swimming in the Olympic-size pool and boating on the river.

Yet no one lives in Pripyat today. The walls are crumbling. Its windows are broken. Its lintels are collapsing. I have to watch my step as I explore its dark, empty buildings. Chairs lie on their backs in the hairdressing salons, surrounded by dusty curlers and broken mirrors. Fluorescent tubes hang down from the supermarket ceiling. The parquet floor of the town hall is ripped up and scattered down the length of a grand, marble staircase. Exercise books litter the floors of school rooms, neat Cyrillic handwriting scoring their pages in blue ink. I find the pools emptied. The seats of sofas in the apartments have dropped to the floor. The beds are rotten. Almost everything is motionless – paused. If something is stirred by a gust of wind, it startles me.

With each new doorway you enter, the lack of people becomes more and more preoccupying. Their absence is the truth that is most present. I've visited other post-human towns – Pompei, Angkor Wat and Machu Picchu – but here, the normality of the place forces your attention on the abnormality of its abandonment. Its structures and accoutrements are so familiar that you know their disuse cannot simply be due to the passing of ages. Pripyat is a place of utter despair because everything here, from the noticeboards that are no longer looked at, to the discarded slide rules in the science classroom, to the shattered piano in the café, is a monument to the capacity of humankind to lose everything it needs, and everything it treasures. We humans, alone on Earth, are powerful enough to create worlds, and then to destroy them.

On 26 April 1986, reactor number 4 of the nearby Vladimir Ilyich Lenin Nuclear Power Plant, known to everyone today as 'Chernobyl', exploded. The explosion was the result of bad planning and human error. The design of Chernobyl's reactors had flaws. The operating staff were not aware of these and, in addition, were careless in their work. Chernobyl exploded because of mistakes – the most human explanation of all.

Four hundred times more radioactive material than that expelled by the Hiroshima and Nagasaki bombs combined was sent over much of Europe on high winds. It fell from the skies in raindrops and snowflakes, entered the soils and waterways of many nations. Ultimately it broke into the food chain. The number of premature deaths caused by the event is still disputed but estimates range into the hundreds of thousands. Many have called Chernobyl the most costly environmental catastrophe in history.

Sadly, this isn't true. Something else has been unfolding, everywhere, across the globe, barely noticeable from day to day for much of the last century. This too is happening as the result of bad planning and human error. Not one hapless accident, but a damaging lack of care and understanding that affects everything we do. It didn't begin with a single explosion. It started silently, before anyone realised it, as a result of causes that are multifarious, global and complex. Its fallout cannot be detected by a single instrument. It has taken hundreds of studies across the world to confirm that it is even happening. Its effects will be far more profound than the contamination of soils and waterways in a few unfortunate countries – it could ultimately lead to the destabilisation and collapse of everything we rely upon.

This is the true tragedy of our time: the spiralling decline of our planet's *biodiversity*. For life to truly thrive on this planet, there must be immense biodiversity. Only when billions of different individual organisms make the most of every resource and opportunity they encounter, and millions of species lead lives that interlock so that they sustain each other, can the planet run efficiently. The greater the biodiversity, the more secure will be all life on Earth, including ourselves. Yet the way we humans are now living on Earth is sending biodiversity into a decline.

We are all culpable but, it has to be said, through no fault of our own. It is only in the last few decades that we have come to understand that every one of us has been born into a human world that was always inherently unsustainable. But now that we do know this, we have a choice to make. We could carry on living our happy lives, raising our families, busying ourselves with the honest pursuits of the modern society that we have built, whilst choosing to disregard the disaster waiting on our doorstep. Or we could change.

This choice is far from straightforward. It is, after all, only human to cling tightly to what we know, and discount or fear what we don't. Every morning, the first thing the people of Pripyat would have seen on drawing back the curtains in their apartments was the giant nuclear power station that would one day destroy their lives. Most of the inhabitants worked there. The remainder relied on those who did for their livelihoods. Many would have understood the dangers of living so close to it, yet I doubt whether any would have chosen to switch the reactors off. Chernobyl had brought them that precious commodity – a comfortable life.

We are all people of Pripyat now. We live our comfortable lives in the shadow of a disaster of our own making. That disaster is being brought about by the very things that allow us to live our comfortable lives. And it is quite natural to carry on in this way until there is a convincing reason not to do so and a very good plan for an alternative. That is why I have written this book.

The natural world is fading. The evidence is all around. It has happened during my lifetime. I have seen it with my own eyes. It will lead to our destruction.

Yet there is still time to switch off the reactor. There *is* a good alternative.

This book is the story of how we came to make this, our greatest mistake, and how, if we act now, we can yet put it right.

PART ONE

My Witness Statement

As I write this, I am 94. I have had the most extraordinary life. It is only now that I appreciate how extraordinary. I have been lucky enough to spend my life exploring the wild places of our planet and making films about the creatures that live there. In doing so, I have travelled widely around the globe. I have experienced the living world first-hand in all its variety and wonder, and witnessed some its greatest spectacles and most gripping dramas.

As a boy, I dreamed, like so many other boys, of travelling to distant wilder places to look at the natural world in its pristine state and even find animals that were new to science. Now, I find it hard to believe that I have managed to spend so much of my life doing exactly that.

1937

World population: 2.3 billion[1]
Carbon in atmosphere: 280 parts per million[2]
Remaining wilderness: 66 per cent[3]

When I was 11 years old, I lived in Leicester in the middle of England. At that time it wasn't unusual for a boy of my age to get on a bicycle, ride off into the countryside and spend a whole day away from home. And that is what I did. Every child explores. Just turning over a stone and looking at the animals beneath is exploring. It never occurred to me to be anything other than fascinated when watching what was going on in the natural world about me.

My elder brother had another view. Leicester had an amateur dramatic society that put on productions of near-professional standards, and although he persuaded me every now and then to join him and speak a couple of lines in walk-on parts, my heart was not in it.

Instead, as soon as the weather was warm enough, I would cycle off to the eastern part of the county where there were rocks full of beautiful and intriguing fossils. They were not, it is true, the bones of dinosaurs. The honey-coloured limestone had been deposited as mud at the bottom of an ancient sea, so no one could expect to find the remains of such land-living monsters in them. Instead I discovered the shells of sea-living creatures – ammonites, some six inches or so across, coiled like rams' horns; others the size of hazelnuts, inside which were tiny scaffolds of calcite that had supported the gills with which the creatures within had breathed. And I knew of no greater thrill than picking up a likely-looking boulder, giving it a smart blow with a hammer and watching it fall apart to reveal one of these marvellous shells, glinting in the sunlight. And I revelled in the thought that the first human eyes to gaze upon it were mine.

I had believed from a very early age that the most important knowledge was that which brought an understanding of how the natural world worked. It was not laws invented by human beings that interested me, but the principles that governed the lives of animals and plants; not the history of kings and queens, or even the different languages that had been developed by different human societies, but the truths that had governed the world around me long before humanity had appeared in it. Why were there so many different kinds of ammonites? Why was this one different from that? Did it live in a different way? Did it live in a different area? I soon discovered that plenty of other people had

asked such questions, and had found a lot of the answers; and that these answers could be put together to form the most marvellous of all stories – the history of life.

The story of the development of life on Earth is for the most part one of slow, steady change. Every creature whose remains I found in the rocks, had spent its entire life being tested by its environment. Those that happened to be better at surviving and reproducing passed on their characteristics. Those that didn't, couldn't. Over billions of years, life forms slowly changed and became more complex, more efficient, often more specialised. And their long story, detail by detail could be deduced from what could be found in the rocks. The Leicestershire limestones had recorded only a tiny moment of it. But more chapters could be found in the specimens that the city's museum had on display. And to find out yet more I decided, when the time came, that I would try to go to university.

There, I learned another truth. This long story of gradual change had been violently interrupted at points. Every hundred million years or so, after all those painstaking selections and improvements, something catastrophic happened – a *mass extinction*.

For different reasons at different times in the Earth's history, there had been a profound, rapid, global change to the environment to which so many species had become so exquisitely adapted. The Earth's life-support machine had stuttered, and the miraculous assemblage of fragile interconnections which held it together had collapsed. Great numbers of species suddenly disappeared, leaving only a few. All that evolution was undone. These monumental extinctions created boundaries in the rocks that you could see if you knew where to look and how to recognise them. Below the boundary there were many different life forms. Above, very few.

Such mass extinctions have happened five times in life's four-billion-year history.[4] Each time, nature has collapsed, leaving just enough survivors to start the process once more. The last time it happened, it is thought that a meteorite over 10 kilometres in diameter struck the Earth's surface with an impact 2 million times more powerful than the largest hydrogen bomb ever tested.[5] It landed in a bed of gypsum, so, some think, it sent sulphur high into the atmosphere to fall across the globe as rain sufficiently acidic to kill vegetation and dissolve the bodies of plankton in the surface waters of the oceans. The dust cloud that arose blocked the light from the Sun to such a degree that it may have reduced the rate of plant growth for several years. Flaming remnants of the blast may have showered back to Earth, causing firestorms across the western hemisphere. The burning world would have added carbon dioxide and smoke to the already polluted air, warming the Earth through a greenhouse effect. And because the meteorite landed on the coast, it initiated colossal tsunamis that swept across the globe, destroying coastal ecosystems and sending marine sand significant distances inland.

It was an event that changed the course of natural history – wiping out three-quarters of all species, including anything on land larger than the size of a domestic dog. It ended the 175-million-year reign of the dinosaurs. Life would have to rebuild.

For 66 million years since then, nature has been at work reconstructing the living world, recreating and refining a new diversity of species. And one of the products of this rebooting of life was humanity.

* * *

Our own evolution is also recorded in the rocks. Fossils of our close ancestors are much rarer than those of ammonites because they first evolved only 2 million years ago. And there is a further difficulty. The remains of land-living animals are not, for the most part, sealed away beneath accumulating sediments as are those of marine creatures. Instead they are smashed by the destructive powers of the baking sun, the driving rain, and frost. But they do exist, and the few remains we have found of our ancestors show that we first evolved in Africa. As we did so, our brains began to increase in size at such a rate as to suggest that we were acquiring one of our most characteristic features – a capacity to develop *cultures* to a unique degree.

To an evolutionary biologist, the term 'culture' describes the information that can be passed from one individual to another by teaching or imitation. Copying the ideas or actions of others seems to us to be easy – but that is because we excel at it. Only a handful of other species show any signs of having a culture. Chimpanzees and bottle-nosed dolphins are two of them. But no other species has anything approaching the capacity for culture that we have.

Culture transformed the way we evolved. It was a new way by which our species became adapted for life on Earth. Whereas other species depended on physical changes over generations, we could produce an idea that brought significant change within a generation. Tricks such as finding the plants that yield water even during a drought, crafting a stone tool for skinning a kill, lighting a fire or cooking a meal, could be passed from one human to another during a single lifetime. It was a new form of inheritance that didn't rely on the genes which an individual received from its parents. So now the pace of our change increased. Our ancestors' brains expanded at extraordinary speed, enabling us to learn, store and spread ideas. But, ultimately, the physical changes in their bodies slowed almost to a halt. By some 200,000 years ago, anatomically modern humans, *Homo sapiens* – people like you and me – had appeared. We have changed physically very little since then. What has changed spectacularly, is our culture.

At the beginning of our existence as a species, our culture was centred upon a lifestyle of hunting and gathering. We were exceptionally good at both. We equipped ourselves with the material products of our culture such as hooks to catch fish and knives to butcher deer. We learned how to control fire for cooking and use stones to grind grain. But, despite our ingenious culture, our lives were not easy. The environment was harsh and, more importantly, unpredictable. The world, in general, was a lot colder than now. The sea level was much lower. Freshwater was harder to find, and global temperatures fluctuated greatly within relatively short periods of time. We may have had bodies and brains very like those we have now, but because the environment was so unstable, it was hard to survive. Data from genetic studies of modern-day humans suggests that in fact, 70,000 years ago, those climatic hazards left us susceptible to events that nearly exterminated us. Our entire species may have been reduced to as few as 20,000 fertile adults.[6] If we were to develop much further we needed a little stability. The retreat of the last glaciers, 11,700 years ago, brought that stability.

* * *

The *Holocene* – the part of the Earth's history that we think of as our time – has been one of the most stable periods in our planet's long history. For 10,000 years, the average global temperature did not vary up or down by more than 1°C.[7] We don't know exactly what produced this stability, but the richness of the living world may well have had something to do with it.

Phytoplankton, microscopic plants floating near the ocean's surface, and vast forests extending right round the globe in the north, locked away a great deal of carbon and so helped to maintain a balanced level of *greenhouse gases* in the atmosphere. Huge herds of grazing animals kept the grasslands rich and productive by fertilising the soils and stimulated new growth by grazing them. Mangrove swamps and coral reefs along the coast provided nurseries for young fish that, when mature, ranged into open waters and enriched the ocean's ecosystems. A dense, multi-layered belt of rainforest around the Equator harnessed the Sun's energy and added moisture and oxygen to the global air currents. And great white expanses of snow and ice at the northern and southern ends of the Earth reflected sunlight back into space, cooling the whole Earth like a gigantic air conditioner.

So the flourishing biodiversity of the Holocene helped to moderate the global temperatures of Earth, and the living world settled into a gentle, reliable annual rhythm – the seasons. On the tropical plains, dry and rainy seasons alternated with clockwork regularity. In Asia and Oceania, the winds changed direction at the same time each year, delivering the monsoon on cue. In northern regions, the temperatures rose above 15°C in March, triggering spring, and then stayed high until October when they dipped and brought autumn.

The Holocene was our Garden of Eden. Its rhythm of seasons was so reliable that it gave our species the opportunities we needed, and we took advantage of them. Almost as soon as the environment stabilised, groups of people living in the Middle East began to abandon gathering plants and hunting animals and took to a completely new way of life. They started to farm. The change was not deliberate. It did not happen by design. The path to agriculture was long, haphazard and accidental, and due more to luck than to foresight.

In the Middle East, the lands had all the characteristics needed for such happy accidents. They lie on the crossroads between three continents – Africa, Asia and Europe – so, for millions of years, species of plants and animals from all three have both passed through and established themselves here. The hillsides and floodplains were colonised by plants such as the wild ancestors of today's wheat, barley, chickpea, peas and lentils – all species that produce seeds so rich in nutriment that they can survive the prolonged dry seasons. Such edible morsels must have attracted people every year. If they were able to gather more seeds than they needed immediately, they doubtless stored them, as some other mammals and birds do, so that they could be eaten during the winter when food is scarce. At some point, the *hunter-gatherers* stopped their wanderings and settled down, secure in the knowledge that their stored seeds would provide them with food when nothing else was easily available.

Wild cattle, goats, sheep and pigs all existed naturally in this region. Initially they must have been taken from the wild, but they too became *domesticated* within a few thousand years of the start of the Holocene. Again, there will have been many intermediate, and doubtless unintentional, steps in the journey from wild to tame. At first, the hunters selected males to kill, and protected breeding females, in order to boost the populations. Evidence for this has been found by scientists studying the bones of animals around ancient village sites. The humans may also have chased off other animal predators or lived without meat entirely for periods of the year to maintain the wild stock. Ultimately, they not only caught but kept animals alive for long periods and began to breed them, inevitably selecting as their stock those individuals that were less aggressive and more tolerant.

With time, all of these developments were enhanced by other innovations – building grain stores, herding, digging irrigation channels, tilling and planting, adding manure. Agriculture had arrived. Perhaps the advent of farming was almost inevitable when a species as intelligent and inventive as ours met a climate as stable as the Holocene's. Certainly, the habit of farming started independently in at least 11 separate regions around the world, gradually developing cultivated strains of a very wide range of crops including familiar ones like potatoes, maize, rice and sugarcane, and domesticated animals such as donkeys, chickens, llamas and bees.

* * *

Farming transformed the relationship between humankind and nature. We were, in a very small way, taming a part of the wild world – controlling our environment to a modest degree. We built walls to protect plants from the wind. We shaded our animals from the Sun by planting trees. Using their manure, we fertilised the land where they grazed. We ensured that our crops flourished in times of drought, keeping them watered by building channels from rivers and lakes. We removed plants that competed with the ones we found useful, and covered whole hillsides with those we particularly favoured.

Both the animals and the plants we selected in this way also began to change. As we protected the grazing animals, they no longer needed to guard against attacks from predators or fight for access to females. We weeded our plots so that our food plants could grow without competition from other species and get all the nitrogen, water and sunlight they needed. They produced larger grains, and bigger fruits and tubers. The animals became more biddable as we took away their need for wariness and aggression. Their ears flopped, their tails curled, they continued to make the yapping, bleating and whining noises of their younger years even when they were mature – perhaps because, in many ways they were eternally youthful, being fed and protected by us, their surrogate parents. And we were also changing from a species that was moulded by nature into one that had the ability to mould other species to match its own requirements.

The farmers' work was hard. They suffered frequent droughts and famine. But eventually they were able to produce more than they needed for their own immediate requirements. Compared to their hunter-gathering neighbours, they were able to raise bigger families. These extra sons and daughters were useful, not only to tend the crops and livestock, but to assist their family in retaining possession of its fields. Farming made land more valuable than it had been in its wild state, and the farmers began to build more permanent shelters to maintain their claims.

The plots belonging to different families inevitably varied in soil type, water availability and aspect. So some crops and herds fared better than others. After feeding the family, the farmers were able to use any surplus to trade. Farming communities came to gather at open markets to barter their wares. They began to exchange food for other assets and for skills. The farmers needed stone, twine, oil and fish. They wanted the products of carpenters, masons and toolmakers, who now for the first time were able to trade for food rather than spending time growing it. As the number of trades increased, the markets developed into towns and then cities in many of the fertile river valleys. As each new valley was settled, some farmers moved to the next in search of fresh fields. Neighbouring tribes of hunter-gatherers, trading with the farming communities, merged with them as they grew, and the practice of farming spread at speed up the rivers into every watershed.

Civilisation had started. It gathered pace with each generation, and with each technical innovation. Water power, steam power, electrification were invented and refined – and eventually all the achievements with which we are familiar today were established. But each generation, in these ever-more-complex societies, was able to develop and progress only because the natural world continued to be stable and could be relied upon to deliver the commodities and the conditions that we needed. The benign environment of the Holocene, and the marvellous biodiversity that guaranteed it, became more important to us than ever.

The town of Pripyat in Ukraine. It was built in the1970s to provide homes for the workers employed in the Soviet nuclear power station at Chernobyl. In April 1986, one of the reactors exploded and the entire population had to be evacuated immediately. The wrecked reactor, seen above on the horizon, has now been enclosed within a giant arched structure of concrete to restrict still dangerous emissions. (© Kieran O'Donovan)

Apartment blocks, built to the latest 1970s design, stand empty, together with dance halls, schools, swimming pools and telephone boxes. All have been abandoned, allowing the forest to return and reclaim its territory. (© Maxym Marusenko/NurPhoto/Getty)

In the studio during a *Zoo Quest to Paraguay* programme. I introduce a six-banded armadillo to the camera, while a two-toed sloth hangs from a tree trunk at the back awaiting its turn in the limelight. (© BBC)

Opposite top: Charles Lagus and I set off for Sierra Leone in 1954. Air navigation then had still not developed sufficiently for overnight flights to West Africa, so we had to spend the first night on the ground in Casablanca. (© David Attenborough)

Opposite bottom: The leader of the hitherto uncontacted Biami in central New Guinea lists the nearby rivers. Counting gestures vary between tribal groups, so the ones he used might reveal which people he traded with. (© David Attenborough)

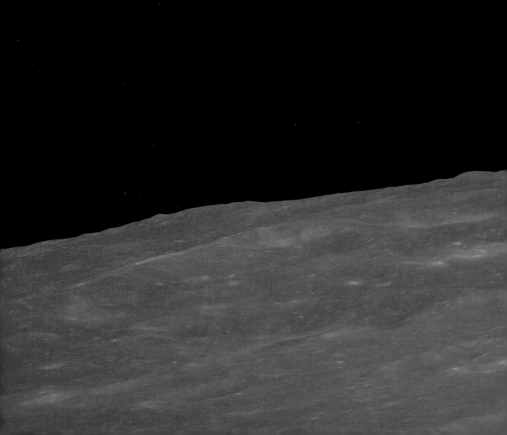

Opposite: Commander Frank Borman in the Apollo 8 spacecraft, which orbited the moon in 1968. (© NASA)

The first sight of Planet Earth, as seen from Apollo 8 – an image that transformed the way in which we perceive our planet and ourselves. (© NASA)

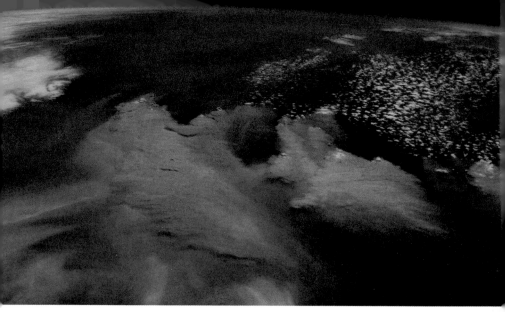

Dense, brown plumes of smoke eclipse patchy white clouds on Australia's southeast coast, as bushfires burn out of control. During the summer of 2019–2020, an estimated 18 million acres went up in smoke, and more than 3 billion animals were killed or displaced. Climate change has been cited as a contributing factor, although many in the Australian Government denied this at the time. (© Geopix/Alamy)

During the filming of *Frozen Planet*, I accompanied scientists from the Norwegian Polar Institute as they darted and anaesthetised polar bears from a helicopter. Research over the years has revealed that the bears are losing weight because of the difficulty of hunting seals on the dwindling sea ice, a trend that, if it continues, is likely to lead to the extinction of the specie. (© BBC)

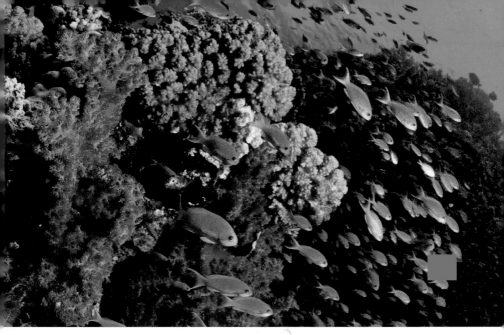

Coral reefs, like this one in the Red Sea, Egypt, are among the most biodiverse habitats on Earth. Yet, while they are rich and complex ecosystems, they are also fragile. At current rates of climate change, some predict that 90 per cent of the world's coral reefs could disappear within decades as the ocean becomes warmer and more acidic. (© Georgette Douwma/naturepl.com)

Coral bleaching, is often caused by warming waters, and is a sign that a reef is under stress. As temperatures rise the coral organisms expel the colourful algae that live within their body tissues. Many of them then die, exposing the white limestone structures they have built for themselves. (© Jurgen Freund/naturepl.com)

Humpback whales, like other large whales, were the targets of commercial whaling fleets in the first half of the twentieth century. Since a ban on hunting, their numbers have recovered from just a few thousand to roughly 80,000 individuals – evidence of how swiftly nature can recover, if given the chance. (© Brandon Cole/naturepl.com)

1954

World population: 2.7 billion
Carbon in atmosphere: 310 parts per million
Remaining wilderness: 64 per cent

After studying the natural sciences at university and doing my national service in the Royal Navy, I joined the infant BBC Television Service. It had started in 1936, the first in the world, using two small studios in Alexandra Palace in north London. It was suspended when the Second World War erupted, but in 1946 it began again, using the same cameras in the same studios. All its programmes were live and in black and white, and they could only be seen in London and Birmingham. My job was to produce non-fiction programmes of all kinds, but as the number and variety of programmes shown each evening increased I started to specialise in natural history.

To start with, we showed animals brought to the studios from the London Zoo. They were placed on a table covered by a doormat and usually handled by one of the Zoo's experts. But that made them look like freaks or oddities. I yearned to let viewers see them in their proper surroundings – in the wild where their varied shapes and colours made sense – and eventually I worked out a way in which I might do that. I made a plan with Jack Lester, the Curator of Reptiles in the London Zoo. He would suggest to the Zoo's Director that he might go to Sierra Leone in West Africa, which he knew well, and that I would go with him with a cameraman to film what he did. After each film sequence showing Jack at work in the wilderness, he would appear live in the studio, show the actual animal that he had caught and explain something about its natural history. It would be excellent publicity for the Zoo, and the BBC would get a new kind of animal programme. We called it *Zoo Quest*. So, in 1954, I set off for Africa with Jack and Charles Lagus, a young cameraman who had worked in the Himalayas and used the lightweight 16mm film camera that we would need.

The first programme was transmitted in December 1954. Unhappily, the day after it appeared Jack was taken to hospital with a disease so serious that it would eventually kill him. There was no way in which he could appear in the studio for the second programme the following week. Only one person could do the job, and that was me. So I was instructed to leave the control gallery from which I had directed the live cameras, and instead stand in the studio grappling with the pythons, monkeys, rare birds and chameleons that the expedition had brought back. So began my career in front of the camera.

The series turned out to be very popular and I started to travel the world making *Zoo Quest* programmes – Guyana, Borneo, New Guinea, Madagascar, Paraguay. Wherever I went, I found wilderness: sparkling coastal seas, vast forests, immense open grasslands. Year after year I explored such places with cameras, recording the wonders of the natural world for the viewers back home. The people who helped us, guiding us through these jungles and deserts couldn't understand how I found it so difficult to locate animals – animals that were plainly obvious to them. It was some time before I acquired the skills that I needed to become reasonably competent at living and working in the wilderness.

The programmes became extremely popular. People had never seen a pangolin before on television. They had never seen a sloth. We showed them the largest lizard, the so-called 'dragon' that lives on Komodo, a small island in central Indonesia, and filmed for the first time birds-of-paradise dancing in the New Guinea forest.

The 1950s were a time of great optimism. The Second World War that had left Europe in ruin was beginning to fade in the memory. The whole world wanted to move on. Technological innovation was booming, making our lives easier, introducing us to new experiences. It felt that nothing would limit our progress. The future was going to be exciting and bring everything we had ever dreamed of. Who was I, travelling the globe with the task of exploring nature, to disagree.

That was before any of us were aware that there were problems.

1960

World population: 3.0 billion
Carbon in atmosphere: 315 parts per million
Remaining wilderness: 62 per cent

If there is a single wilderness of which everyone has a clear mental image, it is the great plains of Africa, with their elephants, rhinos, giraffes and lions. My first visit to the plains was in 1960. While the wildlife I encountered was wonderful, it was the sheer expanse of the wide-open landscapes that captured my attention. The Maasai word 'Serengeti' means 'endless plains'. It's an apt description. You can be in one spot on the Serengeti, and the place appears to be totally empty of animals – and then the next morning, there are one million wildebeest, a quarter of a million zebra, half a million gazelle. A few days after that, and they're gone, over the horizon, out of sight. You'd be forgiven for thinking that these plains were endless, when they can swallow up such immense herds.

At that time, it appeared inconceivable that human beings, a single species, might one day have the power to threaten something as vast as this wilderness. Yet that was exactly the fear of a visionary scientist, Bernhard Grzimek. He was the Director of the Frankfurt Zoo, and had revived it from a wreckage of broken cages and bomb craters after the war. In the 1950s, he became a familiar face on German television, presenting films on African wildlife. His most famous, *Serengeti Shall Not Die*, had won the Academy Award for Documentary in 1959. The film recorded his work to try to map the movements of the wildebeest herds. He and his son Michael, who was a qualified pilot, had used a small plane to follow the herds over the horizon. They charted them as they moved across rivers, through woodlands and over national borders, and in doing so he began to understand the workings of the entire Serengeti ecosystem. It became apparent that the grasses, surprisingly, needed the herbivores as much as the herbivores needed the grasses – without the grazers, the grasses wouldn't be so dominant. They had evolved to withstand being cropped by a million voracious mouths. When the herds' teeth sliced off the leaves close to ground level, the plants used reserves of sustenance in their bases just below ground to regrow. When the herds' hooves broke up the soil and the plants shed their seeds, the next generation of grass was established. When the herds moved on, the grasses were able to regrow quickly, nourished by the piles of manure that the animals had left behind them. What looked like a path of destruction in the herds' wake was in fact an essential stage in the grasses' life cycle. If there

34

were too few grazers, the grasses would disappear shaded by taller plants that would come to dominate in the herds' absence.

It was a tale of inter-dependence characteristic of the discoveries then being made by the emerging science of *ecology*. The task of naming and classifying the world's species that had preoccupied zoologists in the nineteenth century was being superseded by other pursuits. Zoologists were now becoming more specialised. Some studied the workings of animal cells, invisible to the naked eye, using ever-more powerful microscopes and X-rays – a pursuit that in 1953 resulted in the discovery of the structure of DNA, the very essence of inheritance. Others, the ecologists, developed statistical techniques and surveying equipment to study animal communities living together out in the wild. In the 1950s, these ecologists were beginning to make sense of the seeming chaos of the outside world and coming to understand how all life was interconnected in a web of infinite variety, with everything relying on everything else. Animals and plants had a close and sometimes intimate relationship with one another, yet, although tightly interwoven, these ecosystems were not necessarily robust. Even a small knock in the wrong place could put the whole community off balance.

Grzimek knew that this must be true even for an ecosystem as large as the Serengeti. His survey flights soon revealed that it was, in fact, the very size of the plains that prevented that ecosystem from collapsing. Without immense space, the herds could not move over great distances and give the different areas of wild pasture the respite they required between attacks. The grazers would grind the leaves down to their very roots, and ultimately bring about their own starvation. The predators might benefit for a short while as their prey became enfeebled by starvation, but with time they too would die. Without its vast space, the Serengeti ecosystem would lose its balance and collapse.

Motivated by the knowledge that Tanzania and Kenya were about to claim independence and might well give in to demands to turn the plains into farmland, Grzimek, through his films and other activities, gave strength to those anxious to protect the grasslands and save the space for nature. The African states, of their own volition, took visionary action. Tanzania banned human settlement in the section of the Serengeti that lay within its borders – a ruling that caused a great deal of controversy. In Kenya, new reserves were created in the area around the Mara River to preserve the entire route of the Serengeti migration.

The point was made. Nature is far from unlimited. The wild is finite. It needs protecting. And a few years later, that idea became obvious to everyone.

1968

World population: 3.5 billion
Carbon in atmosphere: 323 parts per million
Remaining wilderness: 59 per cent

During the *Zoo Quest* expeditions, I had spent time with people in distant parts of the world who led lives very different from my own and I started to learn more about them and the way they viewed existence. I felt that it would be valuable to bring their lives and perspectives to the home audience, so the emphasis of my overseas filming began to change, and I began to make films showing the lives and customs of the people far from Europe – in Southeast Asia, in the islands of the Western Pacific and in Australia. I become so engaged with these peoples that I decided I ought to know something more of their beliefs and the way they organised their lives. The BBC allowed me to resign from full-time employment as a producer and for the next few years to spend six months in twelve making programmes, and then a similar length of time studying anthropology at the London School of Economics. It seemed a marvellous arrangement – but it didn't last long.

In the 1960s the BBC was given the responsibility of introducing colour television into Britain which till then had been in black and white. This would be done by a new network called BBC2. Its programmes would also explore new styles and subjects. What exactly these were, was not defined; that would be the responsibility of its controller. To anyone interested in broadcasting, such a job was irresistible. At any rate, when I was offered it, I found it so, and in 1965 I abandoned my anthropological studies and returned to the staff of the BBC – and to an executive desk.

So it was that in 1968, four days before Christmas, I was standing at the back of the international control room in the BBC's Television Centre watching pictures being sent back to Earth from the Apollo 8 mission. We all knew that Apollo 8 would be special. For the first time, a crew would leave the Earth's orbit, travel all the way to the Moon, circle it, taking pictures of its far side, never before glimpsed by humankind, and return to Earth. It was to be a dry run for the attempt to land on the Moon which President Kennedy had been determined should happen before the end of the decade.

While the mission's focus was certainly on the Moon, it was pictures of the Earth that unexpectedly captured the crew's, and our, attention. Frank Borman, Jim Lovell and Bill Anders were the first people to move far enough away from the Earth to be able to see the whole planet with the naked eye, and it made a deep impression. Three and a half hours into the flight, Jim Lovell spoke his thoughts to NASA:[8] 'Well, I can see the entire Earth now out of the centre window.' They were stunned. 'Beautiful' was the word that all three kept uttering. Anders rushed to get the mission's still camera and become the first person to take a photograph of the whole Earth. It is a spectacular shot, the Earth upside down, almost filling frame with South America lit by the December summer sun. Yet, this photograph, like all those taken on the mission, remained undeveloped in camera until touchdown. What we were waiting for in television studios across the world was an electronic picture.

As the time of the first scheduled broadcast from the craft approached, more people were tuning in around the world than had ever before watched a single television programme. We were greeted, incredibly, by a good picture of the interior of the capsule. After a few pleasantries, Frank Borman explained that Anders, who was operating the video camera, was waiting for the spacecraft to roll to a position where he could point the lens through the window at the Earth.

'Now we are coming up on the view that we really want you to see,' he said to us all.

But at that moment the picture disappeared. Mission Control at Houston scrambled to tell the crew that the picture was breaking up. We all waited, helpless. After a few minutes of fiddling live on air, we were told that the telephoto lens was the problem. Anders switched to the wide-angle lens, but there was still no picture. 'You don't have a lens-cover on there, do you?' said Houston. 'No,' Borman replied curtly. 'We checked that, as a matter of fact.'

Then the first pictures suddenly appeared on all our screens. A disc was visible in the frame, but the wide lens made it quite small. The bigger problem, however, was exposure. The Earth was just too bright, flooded with light from the Sun. 'It's coming in as a real bright blob on the screen,' reported Houston. 'It's hard to tell what we are looking at.'

'That is the Earth,' said Borman, almost apologetically.

Unable to improve the image, the crew gave us a tour of the interior of the spacecraft. We watched the astronauts having their zero-gravity lunch. Jim Lovell wished his mother a happy birthday. And the transmission ended. 'I hope we can get that other lens fixed,' said Borman.

We had to wait a whole day for the next broadcast to witness another attempt. On 23 December, the global viewership had grown to an estimated one billion people – by far the biggest audience in history. Borman began with a proud announcement: 'Hello Houston, this is Apollo 8. We have the television camera pointed directly at the Earth now.' The crew had no viewfinder, so in fact could not know exactly what was in frame.

'We're getting a darn good look at the corner of it,' said Houston, but then the Earth swiftly swung off and disappeared. The telephoto lens was working at least, but there followed agonising minutes of 'left a bit, right a bit', as the crew, working blind, attempted to point the lens at the Earth while the craft yawed gently at a distance of 180,000 miles.

Yet, even though the Earth was slipping and sliding across the television screen, the fact was that a quarter of humankind was watching itself. You hardly dared blink. *That* was the Earth that held the whole of humanity – apart from the three men in the spacecraft who were taking the picture.

With that one image, at Christmas in 1968, television enabled humankind to understand something that no one before had been able to visualise in such a vivid way, perhaps the most important truth of our times – that our planet is small, isolated and vulnerable. It is the only place we have, the only place where *life* exists as far as we can tell. It is uniquely precious.

The pictures from Apollo 8 had transformed the mindset of the population of the world. As Anders himself said, 'We came all this way to explore the Moon, and the most important thing is that we discovered the Earth.' We had all simultaneously realised that our home was not limitless – there was an edge to our existence.

1971

World population: 3.7 billion
Carbon in atmosphere: 326 parts per million
Remaining wilderness: 58 per cent

When I had accepted the administrative job at the BBC in 1965, I had asked that I be allowed every two or three years to leave my desk for a few weeks and make a programme. That way, I maintained, I would be able to keep up with the ever-changing technology of programme-making. And in 1971, I thought of a possible subject.

Until the early twentieth century, European travellers, venturing beyond their continent into distant unexplored corners of the Earth, had to travel on foot. If the country ahead was totally unknown, they recruited porters to carry all the food, the tents and other equipment that would be needed if they were to be self-sufficient far from civilisation. But, in the twentieth century, the development of the internal combustion engine put a stop to that. Explorers now used Land Rovers and jeeps, light aircraft and even helicopters.

I knew of only one place where great discoveries were still being made by explorers travelling entirely on foot – New Guinea.

The interior of this thousand-mile-long island lying north of Australia is filled with steep mountain ranges covered with tropical forest. Even in the 1970s, there were still patches of it that no outsider had yet entered, and walking with a long line of porters was still the only way that anyone could do so. Such an expedition would surely make a fascinating film.

At the time, the eastern half of New Guinea was administered by Australia. I got in touch with friends in Australian television. They found out that a mining company had asked for permission to go into one of these unknown areas to prospect for minerals. Government policy, however, stipulated that no one was allowed to do such a thing before it had been established whether or not there were any people living there. Aerial photographs had not revealed any huts or other buildings, but there were one or two tiny pinpricks in the carpet of forest that might indicate man-made clearings. None were big enough to allow a helicopter to land. The only way to discover what they were was to send in a patrol on foot. And I together with a camera team could accompany them – if I really wanted to do so.

My plan was simple. The nearest European settlement to the area in question was a small government station called Ambunti on the Sepik, the great river that runs roughly eastwards, parallel to the north coast of the island before emptying into the Pacific. The government officer who would lead the expedition, Laurie

46

Bragge, was based there and he would recruit some porters. We would charter a float plane that would land on the river alongside his station and join him.

It turned out to be the most exhausting journey that I have ever made. Laurie had managed to assemble a hundred porters, but even that was not enough to carry all the food that we would need. We would have to have an air-drop of more supplies after about three weeks. We also had to travel across the grain of the country. Every morning soon after dawn, we started walking, cutting our way through the densest forest I have ever encountered, hauling ourselves up steep muddy slopes to the crest of a ridge and then slithering down the sodden undergrowth on the other side, to wade across a small winding river and then do the same thing, over and over again. At four o'clock every afternoon we stopped, made camp and put up tarpaulins to give us shelter from the drenching rains that would start promptly at five.

After three and a half weeks of this, one of the porters noticed human footprints in the forest on the edge of the patch we had cleared. Someone had been close to our camp the previous night, watching us. We followed the tracks. Night after night, having pitched our tents, we put out gifts – cakes of salt, knives and packets of glass beads. One of the porters was stationed to sit on a tree stump and call out every few minutes, saying that we were friends and were bringing gifts. But it was unlikely that the people we were following, whoever they were, would understand him for there are over a thousand mutually incomprehensible languages spoken in New Guinea. Even small groups had their own distinct language. Night after night we called. Morning after morning, the gifts lay where we had left them.

After three further weeks of walking, our supplies were running low. We made camp and, for the next two days, the porters laboriously cut down huge trees to create a clearing on which a helicopter might drop fresh supplies. The drop was successful and accurate and we set off, the porters once again with reassuringly heavy loads – but not complaining, for we had been on short rations. Four weeks after we had started, we were nearing country that had already been mapped. It seemed the expedition and our film, would have no satisfying conclusion.

And then, one morning, I woke up beneath my tarpaulin and saw outside a group of small men, standing within a couple of yards of me. None of them was more than about a metre and a half tall. They were naked except for a broad belt of bark into which they had pushed a bunch of leaves, at the front and the back. Several had what I later discovered were bats' teeth stuck through holes that they had pierced in the sides of their noses. Hugh, the cameraman, who always slept with his camera within arm's reach fully loaded and ready to shoot, was already recording. The men stared at us, wide-eyed, as though they had never seen our like before. I doubtless did the same. I had never seen anyone like them either.

To my surprise I found that it was not difficult to communicate with them. I tried by gestures to indicate that we were short of food. They pointed to their mouths, nodded and opened their string

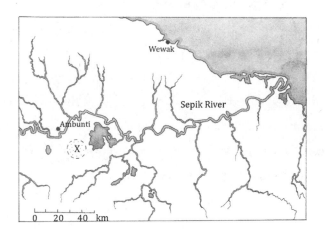

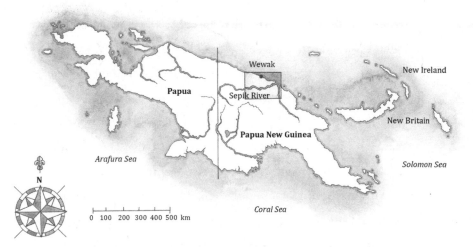

bags to show us roots, probably taro, that they had been gathering. I pointed to cakes of salt we had brought with us. It is used as currency all over New Guinea. They nodded. We had started to trade. Laurie then asked them the names of the nearest rivers. That was more difficult to explain, but they eventually understood what he wanted and they began to list them. How many did they know? They counted them, touching first their fingers one by one, tapping places up their forearm, their elbow, and continuing up the arm and ending on the side of the neck. In fact, Laurie was not particularly interested in the actual names of the rivers, or how many there were. He wanted to know what gestures they used to indicate number. He knew the counting gestures used by other groups in the area, and the ones used by these little people would tell him what trading contacts they might have.

After ten minutes or so, the men started to wave their arms and roll their eyes, indicating that they were going to leave. We waved back in response, trying to invite them to return in the morning with more food. And they left.

The following morning, they reappeared with more roots as we had hoped they would do. We asked if we might see where they were camped and perhaps meet their women and children. After some confusion – or was it perhaps reluctance – they nodded and led us off into the forest. We followed a few yards behind them. It was hard going. The vegetation was very thick. We lost sight of them as we rounded the trunk of a gigantic tree; on the other side, there was no sign of them. They had vanished. We called.

But there was no reply. Were we walking into an ambush? We had no idea. After calling for several minutes, we turned and walked back to camp.

I had had a vision of how all human beings had once lived – in small groups that found all they needed in the natural world around them. The resources they relied upon were self-renewing. They produced little or no waste. They lived sustainably, in balance with their environment in a way that could continue effectively, for ever.

A few days later, I was back in the twentieth century and behind my desk in the Television Centre.

1978

World population: 4.3 billion
Carbon in atmosphere: 335 parts per million
Remaining wilderness: 55 per cent

BBC2 had pioneered one particularly ambitious format – a series of 13 50-minute or one-hour programmes that methodically examined a big and important subject. The first of these was devised to demonstrate the high quality of the new colour system the BBC had adopted by showing the most beautiful and famous paintings, sculptures and buildings produced in Europe over the past thousand years. It was written by the art historian Sir Kenneth Clark, and it took three years to make. Two and a half million viewers watched in Britain. Double that number did so in the United States. It got rave reviews. It was such a success that I immediately commissioned a sequel. This one would examine the history of western science. That in turn was to be followed by a series to mark the bicentenary of the foundation of the United

States, and there would be yet others. But it was very clear to me that the format should also be used to tell the greatest story of all – the history of life itself. It would be the most illuminating series that anyone could wish to make. And I yearned to do so myself. But it couldn't be combined with doing any other job. I had now, however, been an administrator for eight years and that, I thought, was enough. So I decided to leave the BBC once again and then suggest the idea to whoever my successor might be.

In due course that happened. The series was accepted. I called it *Life on Earth*. It took some time to assemble a production team. I wrote the scripts for the 13 episodes more or less in one go. Camera crews were recruited and organised to film at least 600 different animal species in at least 30 countries. I would appear in vision occasionally, to set the scene, to explain complicated theoretical points, introduce new topics or leave frame in one continent and explain in the next frame that we had arrived in another to continue the story. I would have to travel with a crew to a series of different locations. It would take 1.5 million miles of travel to capture the story – two long trips around the globe for me, and the continuous labour of six different camera teams each of whom would be away for months at a time. We would also need a few sequences that were so difficult to get they would be best tackled by camera operators who had special knowledge and skills in filming particular kinds of subjects – oceanic plankton, spiders, hummingbirds, coral fish, bats, and dozens of others. Relating the history of life was the biggest single project I had

ever tackled and would take the next three years of my life. It was an exciting prospect.

One of the key sequences planned for the programme describing the evolution of monkeys and apes concerned the development of the opposable thumb. This is the anatomical characteristic that enables a monkey to grasp a branch – or a human being to wield a tool and eventually hold a pen – an ability that played a crucial part in the rise of our own species and our civilisations. We could have chosen any species of monkey or ape to illustrate the point, but John Sparks, the director of the episode, decided that it would most dramatic to do so by filming gorillas. He had discovered that an extraordinary American biologist, Dian Fossey, had been living with a group of rare mountain gorillas in the central African state of Rwanda, and had so accustomed them to the presence of human beings that even strangers – providing Dian accompanied them – could get quite close to them. He contacted her. The animals with which she worked were seriously endangered. The human population of Rwanda was growing extremely swiftly and the mountain forest in which the gorillas lived was being felled by the local people to make way for cultivated fields. Less than 300 mountain gorillas were left. Their appearance on television might draw the attention of the world to their plight. With that in her mind, she agreed to help us, and in January 1978 we set out for Rwanda.

We landed at Ruhengeri, a tiny airstrip as close as we could get to Dian's camp. From there we would have several hours walking up the volcano's flank to reach the high–altitude forest where Dian lived. We were met by Ian Redmond, a young scientist who was working with Dian. He had very bad news. A young male gorilla that Dian had known since his birth, and was particularly fond of, had been found dead and horribly mutilated. Poachers had shot him. They had cut off his head and his hands to sell to traders who would turn them into souvenirs. Dian was grief-stricken. She was also seriously ill with a lung infection so she had been unable to leave camp. Nonetheless, she would do her best to help us.

The climb to her camp was long and arduous. When at last we arrived, we found her in bed in her cabin, coughing blood. She was clearly seriously ill, but she insisted that she would be well enough to lead us to her gorillas.

The next day she was still very frail so it was Ian who led us into the forest. I had never been in country anything like it. Stunted gnarled trees, wreathed in mist, stood above thickets of giant celery and nettles that reached up to our shoulders, Once we had found the gorillas' tracks, following them through such undergrowth was easy. After an hour or so we could hear crashes ahead of us and we knew we were close. As we moved cautiously forward, Ian started to make a series of loud grunting noises to signal our presence. It was important not to catch them by surprise. If we did the dominant male might charge us. We came to a clearing and Ian called a halt. We must now sit out in the open so that the gorillas could see us. Once they knew we were with Ian, they would be unlikely to take fright.

After a few minutes' rest, we set off again and soon caught up with a family group of them. They were feeding, ripping up the vegetation by the handful. We sat and watched enthralled until, after a few minutes, they got to their feet and leisurely strolled away. We had been accepted, Ian said. Next time, we could film.

The following day, with Ian as our guide, we filmed the gorillas foraging, from a respectful distance. They took virtually no notice of us. Eventually John suggested that I said something directly to camera. explaining what it was like to be sitting near them. We moved slowly towards a group busy feeding, and I cautiously moved closer to them until I thought that they would be visible in the background. I looked back at the camera and spoke.

'There is more meaning and mutual understanding in exchanging a glance with a gorilla,' I said quietly, 'than with any other animal I know. Their sight, their hearing, their sense of smell are so similar to ours that they see the world in much the same way as we do. We live in the same sort of social groups with largely permanent family relationships. They walk around on the ground as we do, though they are immensely more powerful than we are. So if there were ever a possibility of escaping the human condition and living imaginatively in another creature's world, it must be with the gorilla. The male is an enormously powerful creature but he only uses his strength when he is protecting his family and it is very rare that there is violence within the group. So it seems really very unfair that Man should have chosen the gorilla to symbolise everything that is aggressive and violent, when that is the one thing that the gorilla is not – and that we are.'

I wished people to know that these animals were not the brutal wild beasts of legend. They were our cousins and we ought to care for them. The awful truth was that the process of extinction that I had seen as a boy in the rocks was happening right here around me, to animals with which I was familiar – our closest relatives. And we were responsible.

When we found them the next day, they were not far from where we had left them. They had settled on a slope on the far side of a small stream. Martin Saunders set up his camera, Dicky Bird the sound recordist fixed a small radio microphone to my shirt. The time had come, John said, for me to say something about the evolutionary significance of the opposable thumb.

I crept down a slope to a small stream, crossed it and crawled up the opposite slope to a point where I thought that Martin and his camera would be able to see both me and them. John gave me the thumbs-up. But before I could say anything, something landed on my head. I turned and found that a huge female gorilla had emerged from the vegetation immediately behind me and put her hand on my head. She looked straight at me with her deep brown eyes. Then she removed her hand from my head and pulled down my lower lip to look inside my mouth. This was not, I thought, the moment to talk about the evolutionary significance of the opposable thumb. Something then landed on my legs. Two infant gorillas were sitting on my feet and fiddling with my bootlaces.

How long, in terms of minutes and seconds, this interaction continued I have no real idea. It was certainly several minutes. I was in a delirium of happiness. Then the youngsters got bored with my bootlaces and ambled away. Their mother watched them, heaved herself to her feet to lumber after them.

I crept back to the film crew overwhelmed with a feeling of extraordinary privilege.

We had to leave the following morning. As we said goodbye to Dian, she made me promise to try to raise money to help protect the wonderful creatures for which she cared so much. And so I did, the day after we got back to London

* * *

We had filmed the world's biggest primate. I thought now that *Life on Earth* should also include shots of the biggest creature that has ever existed – a whale.

The great whales have been hunted for millennia by brave men in canoes using nothing more than a handheld harpoon. To begin with, the balance of power was with the whales. Not only did they dwarf their human hunters, but they were able to dive within seconds and escape into the depths of the ocean. In the twentieth century, however, that balance tipped dramatically the other way. We invented ways of tracking whales down and stabbing them with harpoons that had explosive heads. Factories, some floating, some on land, were built that were capable of processing several giant carcasses in a day. Whaling had become industrialised. By the time

I was born, 50,000 whales were being killed every year to supply an established market for their oil, their meat and their bones.

The first whales evolved from land-living creatures. The size of terrestrial animals is limited by the mechanical strength of bone: above a certain weight, bone breaks. Aquatic animals, however, are supported by water so whales can grow much bigger than any land animal. And they do. Their nostrils migrated to the tops of their heads, their forelimbs and tails became paddles, and their hind limbs eventually vanished. For tens of millions of years, they were important members of the complex ecosystems of the open ocean, criss-crossing the seas in their hundreds of thousands.

A key problem restricting life in the open ocean is availability of nutrients. Where conditions are right, plants and animals live in the surface waters and, when they die, drift continuously downwards as 'marine snow'. Where nutrients are not freely available, the surface waters of the oceans can be almost sterile. Just as land plants need fertiliser as well as sun and water, so phytoplankton, the photosynthesising foundation of the ocean food web, need nitrogenous compounds in the sunlit surface waters if they are to thrive. There are places in the ocean where the decomposed marine snow is stirred and carried upwards by the currents flowing over submarine mountains and ridges, and here the phytoplankton – and hence fish populations – can flourish. But the rest of the open ocean would remain a vast, blue desert were it not for the whales. They are so big that when they dive to feed in the depths or rise to the surface to breathe, they create a great stirring of the water around them. That helps keep nutrients near the surface. And when they defecate, the waters around them are also greatly enriched. This 'whale pump', as it is often termed, is now recognised as a significant process in maintaining the fertility of the open ocean. Indeed, whales are now thought to be responsible for bringing more important nutrients to surface waters in some parts of the ocean than the outflows of local rivers.[9] The ocean of the Holocene needed its whales to remain productive. In the twentieth century, men killed close to 3 million of them.[10]

Whales cannot withstand such a level of hunting for long. Given the chance, they are very long-lived. Sperm whales can live for 70 years. The females are not sexually mature until they are nine. Their pregnancy lasts for over a year and they give birth only once in every three to five years. As the industrial whalers became more and more efficient, they selected the largest animals when they had the choice, for they were the most profitable. The whales were unable to give birth fast enough to replace their dead.

When we started to plan the filming for *Life on Earth*, no one, as far as we could discover, had ever filmed a living blue whale out in the open ocean. We planned to change that. But in the 1970s their population had been reduced from an estimated 250,000 before industrial whaling began to no more than a few thousand. Distributed widely over the great expanses of the open ocean and still being chased by whalers, they were virtually impossible to find.

Instead, we went in search of humpbacks off Hawaii. We had an additional tool in our kit to help us find them – a hydrophone.

In the late 1960s, an American biologist Roger Payne had turned from recording the ultrasonic sounds of bats to investigating claims from the US Navy that there were songs in the ocean. The Navy had set up listening stations for Soviet submarines and, as well as the signature sound of propellers, they were detecting strange almost musical serenades. Payne discovered that the chief source of these songs were the 5,000 or so humpback whales that were still alive at the time. His recordings revealed that humpback songs are long and complex, and of such low frequency that they can carry for hundreds of kilometres through the water. Humpbacks living in the same part of the ocean, learn their songs from each other. Each song has its own distinct theme on which each individual male will invent his own variations. These change over time. Whales, you might say, have a musical culture.

Payne released his recordings on vinyl discs in the 1970s, and they became hugely popular, transforming the public's perception of whales. Creatures that had been viewed as little more than a source of animal oil now became personalities. Their mournful songs were interpreted as cries for help. In the highly charged political atmosphere of the 1970s, a powerful, shared conscience suddenly stirred. An anti-whaling campaign began with a few passionate supporters and rapidly developed into a mainstream activity. Human beings have pursued animals to extinction many times in our history, but now that pursuit was visible in the shaky, handheld video footage brought back by brave anti-whaling campaigners and it was seen to be no longer acceptable. The surface of the ocean slicked with blood, the butchery in the factories, could not be concealed and the killing of whales turned from a harvest into a crime.

Nobody wanted animals to become extinct. People were beginning to care for the natural world, as they became more aware of it. And television was a way of helping them to do so, all around the world.

* * *

After three years work, *Life on Earth* was transmitted in 1979. It was sold to a hundred territories worldwide and watched by an estimated half a billion people. The series opened with an introduction I called 'The Infinite Variety' – a broad survey of animal and plant diversity, to establish at the outset of the series, that variety is indeed crucial to life. After 11 further chapters expressing the twists and turns in the journey that brought about such variety, the thirteenth and final episode concentrated on just one species – our own.

I did not want to suggest that humanity was in some way separate from the rest of the animal kingdom. We do not have a special place. We are not the preordained and final pinnacle of evolution. We are just another species in the tree of life. Nonetheless we have broken free from many of the constraints that affect all other species. So in the last episode of the series I stood in St Peter's Square in Rome, surrounded by a great crowd of individual *Homo sapiens* from all over the world, and tried to make the point.

'You and I,' I said, 'belong to the most widespread and dominant species of animal on Earth. We live on the ice caps at the poles and in the tropical jungles at the Equator. We have climbed the highest mountain and dived deep into the seas. We have even left the Earth and set foot on the Moon. We are certainly the most numerous large animal. There are something like 4,000 million of us today. And we've reached this position with meteoric speed. It's all happened within the last 2,000 years or so. We seem to have broken loose from the restrictions that have governed the activities and numbers of other animals.'

I was now in my fifties, and there were twice the number of people on the planet as there had been when I was born. Humans had become increasingly separate from the rest of life on Earth, living in a different and unique way. We had eliminated almost all of our predators. Most of our diseases were under control. We had developed ways of producing food to order, and of living in great comfort. Unlike all other species in the history of life on Earth, we were free from the pressures of evolutionary natural selection. Our bodies had not changed significantly in 200,000 years, but our behaviour and our societies had become increasingly detached from the natural environment that surrounds us. There was nothing left to restrict us. Nothing to stop us. Unless we stopped ourselves, we would continue to consume the Earth's physical resources, until we had used them up.

The courageous efforts of Dian Fossey, the successes of the anti-whaling campaign, Peter Scott's rescue of the Hawaiian goose, the reintroduction into the wild of the Arabian oryx, the creation of tiger reserves in India – all the work being done by a growing army of conservationists, passionately raising funds and pushing for policies to protect precious species, would not be enough. And because *Homo sapiens* always wants more, the next stage was inevitable. Whole habitats would soon start to disappear.

1989

World population: 5.1 billion
Carbon in atmosphere: 353 parts per million
Remaining wilderness: 49 per cent

I saw my first orangutan on 24 July 1956, on the third of my *Zoo Quest* journeys. It was a memorable encounter, my first wild great ape – a giant male, a furry red form swaying in the branches, peering down at me with interest and apparently some disdain. The film we took of him was far from perfect. He was half-hidden and silhouetted against the light but, as far as I knew, television had never shown a shot of one in the wild before. Local hunters from the longhouse in which we were staying, halfway up the Mahakam River in eastern Borneo, had found it for us. As we left, one of them took a shot at it with his gun. I turned around, outraged. Why had he done that? Apes like that, he replied, raided the crops that he grew to feed his family. Who was I to tell him that he should not do so?

Rainforests are particularly precious habitats, the most bio-diverse places in the world. More than half of the land-living species on the planet are found in their green depths. They grow in moist, tropical regions where there is an abundance of those two resources that nearly all plants need – freshwater and sunshine. Close to the Equator, the Sun shines for 12 hours each day with such reliability that there are virtually no seasons. Air currents collect water from all over the tropics and drench the forest with up to four metres of rain a year. And the forest also circulates its own water – moisture from a trillion transpiring leaves rising as mists each morning as the Sun warms to full power, only ultimately to fall again as rain.

The supreme suitability of these places for plants results in the greatest and most vigorous competition for space that occurs anywhere on Earth. Giant trees, soaring 40 metres into the air, stretch out their massive branches in all directions to claim the light. Together they create something that is very rare on land – a truly three-dimensional habitat. Beneath a crowning canopy the branches serve as highways to all parts of the forest for those that are unable to fly. Way below, on the dark floor, a tangle of massive roots and tiny threads give stability to the huge trunks. Thousands of other plants support themselves in a multitude of ways. Some rise to claim a place in the sun by climbing up the tree trunks from below. Others, perhaps deposited as seeds by birds, establish them-selves on the massive branches. Still more live close to the ground in the relative darkness, growing slowly on such sustenance as can be derived from a carpet of dead leaves.

And within and throughout this vegetation there are animals. Small species greatly outnumber the large ones. There are numerous invertebrates, small mammals and birds – seed-eaters, bark-gnawers, sapsuckers, flower-lickers, fruit-pickers, leaf-cutters. Their interdependent lives are never-failingly wonderful to the naturalist who tries to disentangle them. Wasps can be found that spend most of their lives within tiny figs, thrips that roll themselves up in flowers, tadpoles that swim in the cups of vase-plants, lizards that disguise themselves with fringes and tatters of skin so that they are completely invisible on a tree trunk until they move. Rainforests are places where evolutionary innovation and experimentation runs wild.

The absence of seasons in the tropics gives a timelessness to the forest that encourages biodiversity. Since the plants are not tied to a climatic calendar, their flowering, fruiting and production of seeds can happen at any time. Some trees fruit more or less continuously. Others grow for months, even years, between sudden flurries of blossoming and fruit production. So pollination, fruit-eating and collecting seeds are not seasonal activities in the rainforest as they are in the forests to the north and south. Food is available the year round, a harvest that is exploited by dozens of different species from scores of different animal groups. Most of the millions of species exist in small numbers and have limited range and many have become highly specialised. One species of insect may live on just one species of plant, perched on one species of tree. The result is a baffling complexity of interconnected relationships – every species a critical component of the whole.

The orangutan that so haunted my memory is an example. The species is widely dispersed in the forests of Borneo and Sumatra, but it performs a crucial role in the seed dispersal of many kinds of canopy trees. Orangutan mothers spend ten years with their single babies, teaching them when and how to collect dozens of different fruits. Being large animals and almost entirely vegetarian, they consume a great deal each day and have to travel continuously in search of crops of ripe fruit. They either spit out the seeds on the spot or carry them in their stomachs for days before dropping them, together within lumps of fertiliser, several kilometres away. Both methods improve the chances of seed germination and in some cases are essential for it to happen.

It is the astonishing variety of tree species in rainforests that underpins their great biodiversity. It is also the characteristic that we are removing. I have visited the forests of Southeast Asia many times for various programmes over the years. Beginning in the 1960s, Malaysia, then Indonesia, began to replace the dizzying diversity of their rainforest trees with just one kind – the oil palm. There were 2 million hectares of oil palm plantation in Malaysia by the time I visited it in 1989 for a series called *Trials of Life*. I remember travelling along a river searching for proboscis monkeys. We were surrounded by the familiar curtain of green, with birds erupting from the foliage every minute or so. Perhaps – I allowed myself to believe – all was well. But on flying back over the area, I saw the forest for what it was – a strip about half a mile wide fringing the water, a forest so narrow and exposed that it would

undoubtedly be degrading each day. Beyond it, and stretching for as far as I could now see from the air, there was nothing but a single species of tree – oil palms in regimented rows.

The disappearance of this rich and remarkable forest has been extremely difficult to accept. The Southeast Asians were simply doing what we in Europe and North America had already done. Satellite shots of either continent today show that the landscape now consists of small islands of dark green forest, separated by vast tracts of cultivated fields. The truth is, there has always been a double incentive to cut down forests. People benefit from the timber, and then benefit again from farming the land that has been exposed. Little wonder that Homo sapiens is such a determined and effective destroyer of forests. It has been estimated that we now have three trillion fewer trees across the world than at the start of human civilisation.[11] What is happening today is just the latest chapter in a process of global deforestation that has been operating for millennia.

Now it is the turn of the rainforests. And as with everything in the latter half of the twentieth century – the latter half of my life – we are working on a scale and at a speed that increases by the year. Half of the world's rainforests have already gone. Borneo's population of orangutan cannot live without the forest, and it has been reduced by two-thirds since I first saw one just over 60 years ago.[12] Orangutans are still easy to find and film, not because they are abundant, but because so many of them now live in sanctuaries and rehabilitation centres, cared for by conservationists alarmed by the pace of the loss they see about them.

We cannot continue to cut down rainforests for ever, and anything that we can't do for ever is, by definition, unsustainable. If we do things that are unsustainable, the damage accumulates to a point when ultimately the whole system collapses. No habitat, no matter how big, is secure.

1997

World population: 5.9 billion
Carbon in atmosphere: 360 parts per million
Remaining wilderness: 46 per cent

The largest habitat of all is the ocean. It covers over 70 per cent of the Earth's surface but, because of its great depths, it accounts for 97 per cent of our planet's inhabitable space. Life on Earth almost certainly began there, probably as microbes living around jets of hot water discharging from vents in the ocean floor, several kilometres below the surface. For 3 billion years, natural selection worked on such single, simple, isolated cells, refining their internal workings. It took 1.5 billion years for cells to reach a structural complexity comparable with that of the cells of which we are made, and a further 1.5 billion before such cells clumped together and worked in a coordinated way as they do in a multicellular organism.[13]

The early marine microbes had metabolisms that released methane as a by-product. It bubbled to the surface and slowly changed the Earth's atmosphere. Earth was a much cooler place at the time. Methane is a greenhouse gas 25 times more potent than carbon dioxide, and its presence in the atmosphere caused the planet to begin to warm, so helping life to proliferate.

Later, microscopic organisms called cyanobacteria began to photosynthesise, using energy from the Sun's rays to build their tissues. The exhaust gas of the process – oxygen – caused a revolution. It became the standard fuel for a much more efficient way of extracting energy from food, and so paved the way for the establishment of all complex life. Cyanobacteria still constitute a significant part of the phytoplankton that floats today in the upper levels of the ocean, You and I, and all the animals with which we share the land, are all ultimately descended from marine creatures. We owe the ocean everything.

In the late 1990s, film-makers in the BBC's Natural History Unit proposed making a series devoted entirely to life in the sea. They called it *The Blue Planet.* The seas are the most difficult and expensive of all environments in which to film and about the hardest places of all in which to record animal behaviour. Bad weather, poor water visibility and difficulty in simply finding animals in the vast three-dimensional expanses of the ocean can ruin any filming day. But the ocean also offered great opportunities for new, startling perspectives on the natural world. The first to show them on television was a Viennese biologist called Hans Hass who, accompanied by his wife Lotte, filmed in the Red Sea. He was followed by Captain Cousteau, who had invented the demand valve, the device that is

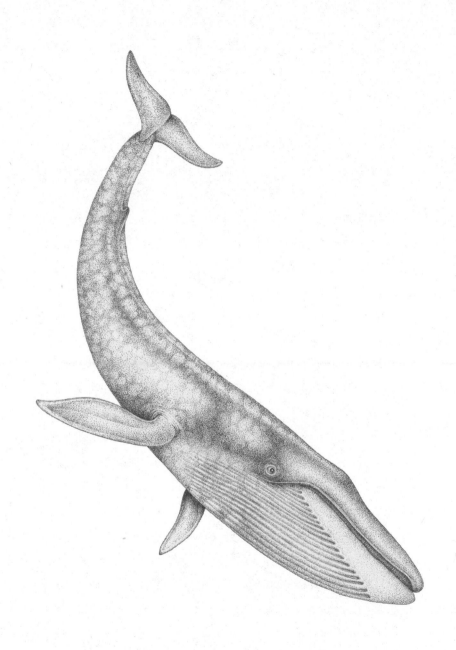

still the essential mechanism enabling human swimmers to breathe underwater. Year after year after year, he indefatigably filmed in oceans around the world. Even after the work of these pioneers, however, the immense variety of life in the sea, so much greater than that which exists on land, had hardly been seen.

The Blue Planet took nearly five years to make and involved almost 200 filming locations. Specialist underwater camera operators recorded cuttlefish courting on the coral reefs, sea otters diving for shellfish in underwater forests of kelp, hermit crabs battling over vacant shells, hammerhead sharks assembling in hundreds to breed off a seamount in the Pacific, and, perhaps most difficult and remarkable of all, sailfish and bluefin tuna hunting in the open ocean. Deepwater craft were used to look for new species on the abyssal plains and watch the carcass of a grey whale being torn apart by hagfish. My contribution was to provide the commentaries.

One team using a microlight aircraft worked for three years to get shots of a blue whale cruising in the open ocean. That sequence opened the series. Here at last was the biggest animal that had ever existed on our planet, hardly ever seen alive and about which we knew almost nothing. But perhaps the great triumph of *The Blue Planet* were the baitball sequences – natural dramas as spectacular as any to be found on the Serengeti. Tuna sweep around the baitfish, penning them against the surface, swimming around them to drive them into a tight, panicked ball. Then they attack, shooting through the ball at lightning speed and from all angles. Ranks of sharks and dolphins charge through the frothing sea to join the fray. Dolphins tackle it from below, encircling the ball with a curtain of bubbles that condenses the baitball still further. Then, just when you thought the commotion would fade, gannets arrive and dive into it from above, slicing through the water to grab beakfuls of fish. And finally, a whale may appear to scoop up the remaining bait in its giant bucket mouth.

Baitball frenzies like these must occur thousands of times a day across the ocean, yet no one had ever seen them before from underwater. They were the most difficult of all the natural events to predict, and therefore to capture on film. In a sense the crew were doing just what the tuna, dolphins, sharks and gannets were doing – waiting for the sudden appearance of an ephemeral 'hotspot' – a great cloud of plankton, feeding on surges of nutrients that rise from the deep on upwelling currents. Such blooms attract huge shoals of smaller fish from hundreds of kilometres away. Once the baitfish among them are present in sufficient density, the predators strike, and in a moment, the ocean becomes a frenzy of action. The camera teams trying to film this event were always playing catch-up – scanning the horizon for diving birds or purposeful pods of dolphins. The *Blue Planet* crews between them spent 400 days without seeing a sign of such an event. And on the few days when the sea did come to life, they had to get alongside the shifting site, and dive beneath the baitball before it was reduced to nothing. It was a high-risk operation. But when it succeeded, it produced unrivalled drama.

Large, commercial fleets first ventured into international waters in the 1950s. Legally, they were working no man's land, places where you could catch as much as you wanted without any restrictions. At first, fishing in largely unexploited seas, the catches were rich. But within only a few years, in any one area, the nets being pulled in were almost empty. So the fleets moved on. After all, wasn't the ocean vast, and virtually unlimited? Checking the data of catches over the years, you can see how one patch of ocean after another was virtually cleared of its fish stocks. By the mid 1970s, the only really fruitful areas were off eastern Australia, southern Africa, eastern North America and in the Southern Ocean.[14] By the start of the 1980s, fishing globally had become so unrewarding that countries with big fleets had to support them with financial subsidies – in effect, paying the fleets to *overfish*.[15] By the end of the twentieth century, mankind had removed 90 per cent of the large fish from all the oceans of the world.

Targeting the seas' largest, most valuable fish is exceptionally damaging. It not only removes the fish at the top of the food chain, such as tuna and swordfish, it also removes the biggest specimens within a population – the largest cod, the biggest snappers. In fish populations, size matters. Most open-water fish grow throughout their lives. The reproductive potential of a female fish is related to her bulk. Large mothers produce disproportionately more eggs. So, by removing all the fish over a certain size, we remove its most effective breeders and soon populations collapse. In heavily fished areas, there are no longer any big fish.

This hunt for fish is a game of cat and mouse that has been refined by generations of fishing communities along the world's coasts. As always, with our unequalled ability to solve problems, we have invented a huge variety of ways to catch fish. Craft have been adapted to particular seas and weathers, and navigation equipment devised, from simple maps to marine chronometers that maintain their accuracy even when tossed about in the heaviest of seas. Predictions as to where hotspots of marine life will appear may draw on the memories of old fishermen or deploy high-tech echo sounders. In the pursuit of fish, we have developed nets that are pushed through the water, nets that drift on the currents, nets that surround a shoal and are then drawn inwards at their base, nets that are cast onto the sea from above and nets that sink and scrape up the seabed. We've measured the depth of the whole ocean, charting its hidden seamounts and continental shelves so that we know where to wait. We work from dinghies and canoes and ships that can spend months at sea, laying walls of net over miles of ocean, taking hundreds of tonnes of fish in a single haul.

We have become too skilled at fishing. And we have done so, not gradually, but – as with whaling and the destruction of rain-forests – suddenly. Exponential gains are characteristic of cultural evolution. Invention accumulates. If you combine the diesel engine, GPS, and the echo sounder, the opportunities they create are not just added to one another, they are multiplied. But the ability of fish to reproduce is limited. As a consequence, we have now over-fished many of our coastal waters.

Taking whole populations of fish from the open ocean is a reckless practice. Ocean food chains operate very differently from those on the land. Chains there may be only three links long – grass to wildebeest to lion. The ocean routinely has chains with four, five and more links. Microscopic phytoplankton are eaten by barely visible zooplankton; that in turn is eaten by small fry, which are then taken by a series of fish of increasing size with bigger and bigger mouths. This extended chain is what we witness at a baitball, and it is self-sustaining and self-regulating. If one kind of mid-size fish disappears because we enjoy them on the plate, those below them on the food chain may become overabundant, and those above may starve because they themselves cannot eat the plankton. The short-lived, finely balanced bursts of life at the hotspots become rarer. Nutrients drop from the waters near the surface of the ocean and tumble down to remain in the gloom below – a net loss to the surface community for millennia. When the hotspots start to diminish, the open ocean starts to die.

The truth is that, over time, we have been forced by our growing numbers to become increasingly efficient catchers of fish. With every year, not only do we have more mouths to feed but there are fewer fish to be caught. Records and reports even from just beyond living memory, back in the nineteenth century and the start of the twentieth, describe an ocean that we wouldn't recognise. Old photographs show people thigh-deep in salmon. Reports from New England tell of fish shoals so vast and so close to shore, that locals would wade in to take them with their table forks. In Scotland, fishermen would haul in a line of 400 hooks and find flatfish on almost all of them.[16] Our not so distant ancestors fished with nothing more complex than hooks and nets of cotton. We now struggle to catch something edible, with technology that would take their breath away.

There are fewer fish in the sea today. We don't realise that this is so because of a phenomenon called the *shifting baseline syndrome*. Each generation defines the normal by what it experiences. We judge what the sea can provide by the fish populations we know today, not knowing what those populations once were. We expect less and less from the ocean because we have never known for ourselves what riches it once provided and what it could again.

* * *

Meanwhile, marine life was also unravelling in the shallows. In 1998, a *Blue Planet* film crew stumbled upon an event not widely known about at the time – coral reefs were losing their normal, delicate colours and turning white. When you first see this, you may think that it is beautiful – the pure white branches, feathers and fronds look like complex marble sculptures – but you soon realise that it is in fact tragic. What you are looking at are skeletons – skeletons of dead creatures.

Coral reefs are built by tiny animals called polyps, related to jellyfish. They have simple bodies consisting of little more than a stalk containing a stomach with a ring of tentacles on top surrounding a mouth. The tentacles have stinging cells which stab passing microscopic prey and convey it into the mouth, which then closes while the polyp digests its capture before reopening for its next meal. These coral polyps build walls of calcium carbonate to protect their soft bodies from hungry predators. Eventually, they become huge stony structures, each species creating its own architectural form. Growing together, they build up into great reefs. The largest of them all, the Great Barrier Reef off north-eastern Australia, is visible from space.

Visiting a coral reef is a fundamentally different wildlife encounter from anything I have known on land. From the first moment that you dive in, you are no longer a prisoner of gravity. You can move in any direction with a flick of one of the fins on your feet. Beneath you stretches a many-hued expanse of coral as grand and varied as a city seen from the air and disappearing into the blue. As you focus, you see that it is populated by a cast of the most extraordinary characters – multi-coloured fish, tiny octopuses, sea anemones, lobsters, crabs and transparent shrimps and all sorts of things that you had no idea ever existed. They are all fantastically beautiful and all, except those right beside you, completely unconcerned by your presence. You float above them, transfixed. If they do look at you, and you stay still, they may approach and even nibble your gloves.

Coral reefs rival rainforests in terms of their biodiversity. They too exist in three dimensions, and that brings the same abundance of opportunities for life as you find in the jungle. But their inhabitants are far more colourful and visible. Spend weeks in a rainforest as I have done, and you begin to seek out parrots and flowers just to experience a colour other than a shade of green. On a reef a whole community of small fish, shrimps, sea urchins, sponges and shell-less tentacle-cloaked molluscs, libellously called sea-slugs, look as if they have been decorated by imaginative schoolchildren in shades of pink, orange, purple, red and yellow.

The colours of the corals come not from the polyps but from symbiotic algae living within their tissues called zooxanthellae. These are able to photosynthesise like other plants. So, as a partnership, the coral polyps and their algal tenants get the benefit of being both plant and animal. During the day, the joint enterprise bathes in the sunlight, the algae using the light to create sugars which supply the polyps with up to 90 per cent of the energy they need. At night, the polyps continue to collect prey. From these meals their algal partners extract the nourishment they require to do their job, and the polyps continue to build their calcium carbonate walls upwards and outwards, so maintaining the colony's position in the sunlight. It is a mutually advantageous relationship that has transformed warm shallow seas, poor in nutrients, into oases of life. But it is one that is precariously balanced.

The bleaching that the *Blue Planet* crews encountered was happening because the corals were becoming stressed and ejecting their algae, exposing the bone-white of their calcium carbonate skeletons. Without their algae, the polyps diminish. Seaweeds begin to colonise the site, smothering the coral skeletons and the reef then turns, with alarming speed, from wonderland to wasteland.

At first, the cause of this bleaching was a mystery. It took a while for scientists to discover that bleaching often occurred where the ocean was rapidly warming. For some time, climatologists had warned that the planet would get warmer if we continued to burn fossil fuels, so adding carbon dioxide and other greenhouse gases to the atmosphere. These gases were known to trap the Sun's energy near to the Earth's surface, heating the planet in a phenomenon called the greenhouse effect. A radical change in the level of atmospheric carbon was a feature of all five mass extinctions in the Earth's history and a major factor in the most comprehensive annihilation of species – the Permian extinction, 252 million years ago. The exact cause of that change is disputed,[17] but we do know that one of the longest and most extensive volcanic events in Earth's history had been growing in strength over a period of a million years, covering what today is Siberia with 2 million square kilometres of lava. This lava may have spread through the existing rocks and reached vast beds of coal, igniting them and discharging sufficient carbon dioxide into the atmosphere to raise the temperature of Earth 6°C above today's average, and increasing the acidity of the entire ocean. The warming of the ocean put all marine systems under stress and, as the waters became more acidic, marine species with calcium carbonate shells – such as corals and much of the phytoplankton – simply dissolved. The collapse of the entire ecosystem was then inevitable. Ninety-six per cent of the marine species on Earth disappeared.

The first phase of a similar ocean death was unfolding while *The Blue Planet* was being filmed in the 1990s. It was an appalling demonstration that we now had the capacity to exterminate living creatures on a vast scale. Furthermore, we were doing so without even entering the sea. This was not like destroying a rainforest. It took hard work to remove the trees. Here, we were damaging distant ecosystems across the world without even visiting them – by changing the ocean's temperature and chemistry with the fallout from our activities thousands of miles away.

It took a million years of unprecedented volcanic activity during the Permian to poison the ocean. We have begun to do so again in less than two hundred. By burning fossil fuels, we are releasing the carbon dioxide captured by prehistoric plants over millions of years in a few decades. The living world has never been able to deal with significant increases of carbon in the atmosphere. Our addiction to coal, oil and gas was on course to knock our environment from its benign, level setting and trigger something similar to a mass extinction.

Yet, until the 1990s, there was little solid evidence for this approaching catastrophe above water. While the ocean was warming, the global air temperature had been relatively stable. The inference was shocking – the air temperature was not changing because the ocean itself was absorbing much of the excess heat of global warming and that was masking the impact that we were making. At some point soon, that would stop. The bleaching corals were like canaries in a coal mine, warning us of a coming explosion. It was the first unmistakable indication to me that the Earth was becoming unbalanced.

2011

World population: 7.0 billion
Carbon in atmosphere: 391 parts per million
Remaining wilderness: 39 per cent

The great wildernesses at both ends of the Earth, in the Arctic and the Antarctic, became the subject for the next major series in which I was involved – *Frozen Planet*. Already in 2011 the world was 0.8°C warmer on average than it was when I was born. That is a speed of change that exceeds any that has happened in the last 10,000 years.

I had visited the polar regions several time over many decades. They have scenery unlike anything else on Earth, and are home to species that have become adapted to a life at the extremes of possibility. But that world was now changing. We realised that the Arctic summers were lengthening. The thaws were starting earlier and the freezes coming later. Camera teams arrived at locations expecting to find expanses of sea ice and found open water. Islands that only a few years earlier had been permanently surrounded by sea ice could now be reached by boat. Satellite images showed that the extent of the summer sea ice in the Arctic had shrunk by 30 per cent in 30 years. Glaciers in many parts of the world were retreating at the fastest rates on record.[18]

And the summer thaw was speeding up. As the air temperature rises and the waters lapping at the edge of ice floes warm, the ice melts faster. As the ice melts, so the whiteness at the two ends of planet Earth shrinks. The dark seas now absorb more of the Sun's heat, creating a positive feedback and speeding the thaw still further. The last time the Earth was as warm as it is now, there was far less ice than there is today. The thaw has a lag – a slow start. But once it gets going, it will be impossible to halt.

Our planet needs ice. Algae grow on the underside of sea ice, sustained by the rays of sunlight that pass through it. The algae are grazed by invertebrates and small fish. They in turn are the basis of the food chains in both the Arctic and Antarctic, some of the most productive seas in the world, providing sustenance to whales, seals, bears, penguins and many other bird species. We too benefit from this chilly productivity. Each year millions of tonnes of fish are caught in both the far north and far south, and sent to markets all over the world.

Warmer summers in the polar regions lead to longer periods without sea ice. For the polar bear, which relies on the northern sea ice as a platform from which to hunt seals, this is devastating. During the summer, they wander idly about the Arctic beaches, sustained by their fat reserves, waiting for the ice to return. As the ice-free period lengthened, scientists detected a worrying trend.

Pregnant females, drained of their reserves, were now giving birth to smaller cubs. It is quite possible that one year the summer will be just that little bit longer, and the cubs born that year will be so small that they cannot survive their first polar winter. That whole population of polar bears will then crash.

Tipping points like this abound in the complex systems of nature. A threshold is reached, often with little warning. It triggers sudden, radical changes that stabilise at a new, altered state. Reversing that tip may be impossible – too much may have been lost, too many components may have been destabilised. The only way to avoid such catastrophe is to watch for warning signs such as the diminishing size of polar bear cubs, recognise them for what they are, and act swiftly.

Further along the Arctic coast of Russia, there is another such sign. Walruses live largely on clams that grow on a few particular patches of the seafloor in the Arctic. In between fishing sessions, they haul themselves out onto the sea ice to rest. But those resting places have now melted away. Instead they have to swim to the beaches on distant coasts. There are only a few suitable places. So two-thirds of the population of Pacific walrus, tens of thousands of them, now assemble on one single beach. Crushingly over-crowded, some clamber up slopes and find themselves at the tops of cliffs. Out of water, their eyesight is very poor but the smell of the sea lying below them at the foot of the cliff is unmistakable. So they try to reach it by the shortest route. The vision of a three-tonne walrus tumbling to its death is not easily forgotten. You don't have to be a naturalist to know that something has gone catastrophically wrong.

2020

World population: 7.8 billion
Carbon in atmosphere: 415 parts per million
Remaining wilderness: 35 per cent

~~~~~~

Our impact is now truly global. Our blind assault on the planet is changing the very fundamentals of the living world. This is now the status of our planet in the year 2020.[19]

We are extracting over 80 million tonnes of seafood from the oceans each year and have reduced 30 per cent of fish stocks to critical levels.[20] Almost all the large oceanic fish have been removed.

We have lost about half of the world's shallow-water corals and major bleachings are occurring almost every year.

Our coastal developments and seafood farming projects have now reduced the extent of mangroves and seagrass beds by more than 30 per cent.

Our plastic debris has been found throughout the ocean, from the surface waters to the deepest trenches. There are currently

1.8 trillion plastic fragments drifting in a monstrous garbage patch in the northern Pacific, where currents cause the surface waters to circulate. Four other garbage patches are forming on similar gyres elsewhere in the oceans.

Plastic is invading oceanic food chains and over 90 per cent of seabirds have plastic fragments in their stomachs. Aldabra is a nature reserve which very few people are permitted to visit. When I landed on the island in 1983, while making *The Living Planet*, the only flotsam on the beaches worthy of mention were the giant nuts of the coco de mer palm tree. Recently another film crew visited the island. They found humanity's rubbish on every part of the beaches. Giant tortoises that live on the island, some over a century old, now have to clamber over plastic bottles, oil cans, buckets, nylon nets and rubber.

No beach on the planet is free of our waste.

Freshwater systems are as threatened as marine. We have interrupted the free flow of almost all the world's sizeable rivers with over 50,000 large dams. Dams can also change the temperature of the water, drastically altering the timing of fish migrations and their breeding events.

We not only use rivers as dumping grounds to remove our litter, but load them with the fertilisers, pesticides and industrial chemicals that we spread on the lands they drain. Many are now the most polluted parts of the environment to be found anywhere on the globe. We take their water and use it to irrigate our crops, and reduce their levels so severely that some of them, at some point in the year, no longer reach the sea.

We build on flood plains and around river mouths, and drain the wetlands to such an extent that their total area is now only half of what it was when I was born.

Our assault on freshwater systems has reduced the animals and plants that live in them more severely than those in any other habitat. Globally, we have reduced the size of their animal populations by over 80 per cent. The Mekong River in Southeast Asia, for example, supplies a quarter of all freshwater fish caught around the globe, and provides 60 million people with valuable protein. Yet a combination of damming, over-extraction, pollution and overfishing has led to a diminishing catch, year by year, not just in volume, but in terms of the size of the fish. In recent years, some fishermen have had to use mosquito nets in order to catch something edible.

Currently we cut down over 15 billion trees each year. The world's rainforests have been reduced by half. The top driver of continuing deforestation, which doubles that of the next three greatest cases combined, is beef production. Brazil alone devotes 170 million hectares of its land, an area seven times the size of the United Kingdom, to cattle pasture. Much of that area was once rainforest. The second driver is soy. Growing soy uses some 131 million hectares, much of it in South America. Over 70 per cent of this soy is used to feed livestock being raised for meat. Third is the 21 million hectares of oil palm plantations, mostly in Southeast Asia.[21]

The forests that still remain are severely fragmented having been intersected by roads, farms and plantations. In 70 per cent of them, the edge of their tree cover is no more than a kilometre away at any point. Few deep, dark forests are left.

Insect numbers, globally, have dropped by a quarter in just 30 years. In places where pesticides are heavily used, this percentage is even higher. Recent studies have shown that Germany has lost 75 per cent of the mass of its flying insects, and Puerto Rico has lost almost 90 per cent of the mass of the insects and spiders living in the canopy The insects are by far the most diverse group of all living species. Many are pollinators, essential links in numerous food chains. Others are hunters and are the dominant factors in preventing populations of plant-eating insects from becoming plagues.[22]

Half of the fertile land on Earth is now farmed. More often than not, we have abused it. We overload it with nitrates and phosphates, overgraze it, burn it, overburden it with unsuitable varieties of crops, and spray it with pesticides so killing the soil invertebrates that bring it to life. Many soils are losing their topsoil and changing from rich ecosystems brimming with fungi, worms, specialist bacteria and a host of other microscopic organisms, into hard, sterile and empty ground. Rainwater runs off it as it does off a pavement and so contributes to the excessive floods that now so frequently submerge the heartlands of many nations that practise industrial farming.

Seventy per cent of the mass of birds on this planet today are domesticated. The vast majority are chickens. Globally, we eat 50 billion of them each year. Twenty-three billion chickens are alive at any one moment. Many of these are fed on soy-based feed derived from deforested land.

Even more startling is the fact that 96 per cent of the mass of all the mammals on Earth is made up of our bodies and those of the animals that we raise to eat. Our own mass accounts for one third of the total. Our domestic mammals – chiefly cows, pigs and sheep – make up just over 60 per cent. The remainder – all the wild mammals, from mice to elephants and whales – account for just 4 per cent.[23]

* * *

Since the 1950s, on average, wild animal populations have more than halved. When I look back at my earlier films now, I realise that, although I felt I was out there in the wild, wandering through a pristine natural world, that was an illusion. Those forests and plains and seas were already emptying. Many of the larger animals were already rare. A shifting baseline has distorted our perception of all life on Earth. We have forgotten that once there were temperate forests that would take days to traverse, herds of bison that would take four hours to pass, and flocks of birds so vast and dense that they darkened the skies. Those things were normal only a few lifetimes ago. Not any more. We have become accustomed to an impoverished planet.

We have replaced the wild with the tame. We regard the Earth as *our* planet, run by humankind for humankind. There is little left for the rest of the living world. The truly wild world – that non-human world – has gone. We have overrun the Earth.

I have spent the last few years speaking about this wherever I can – the United Nations, the International Monetary Fund, the World

Economic Forum, to financiers in London and the festival-goers in Glastonbury. I wish I wasn't involved in this struggle, because I wish the struggle wasn't necessary. But I've had unbelievable luck and good fortune in my life. I would certainly feel very guilty if, having realised what the dangers are, I decided to ignore them.

I have to remind myself of the dreadful things that humanity has done to the planet in my lifetime. After all, the Sun still comes up each morning, and the newspaper drops through the letterbox. But I think about it most days to some degree. Are we, like those poor people in Pripyat, sleepwalking into a catastrophe?

PART TWO

What Lies Ahead

I fear for those who will bear witness to the next 90 years, if we continue living as we are doing at present. The latest in scientific understanding[1] suggests that the living world is on course to tip and collapse. Indeed, it has already begun to do so, and is expected to continue with increasing speed, such that the effects of its decline will become greater in scale and more impactful as they follow one after the other. Everything we have come to rely upon – all the services that the Earth's environment has always provided us for free – could begin to falter or fail entirely. The forecast catastrophe would be immeasurably more destructive than Chernobyl or anything we have experienced to date. It would bring far more than flooded real estate, stronger hurricanes and summer wildfires. It would irreversibly reduce the quality of life of everyone who lives through it, and of the generations that follow. When the global ecological breakdown does finally settle and we reach a new equilibrium, humankind, for as long as it continues to exist on this Earth, might be living on a permanently poorer planet.

The devastating scale of the catastrophe now forecast by mainstream environmental science is a direct result of the way we are currently treating the planet. Beginning in the 1950s after the war, our species entered what has been termed the *Great Acceleration*. Measures of impact and change across a host of parameters demonstrate a strikingly similar pattern when plotted on a graph against time. The trends in our activities can be expressed in terms of *gross domestic product* (GDP), energy use, water use, the building of dams, the spread of telecommunications, tourism, the spread of farmland. You can analyse the change in the environment in many ways – by measuring the rise of carbon dioxide or nitrous oxide or methane in the atmosphere, the surface temperature, *ocean acidification*, loss of fish populations, tropical forest loss. But whatever you measure, the line on the graph will appear to be much the same. From the mid century, it will show a sharply accelerating rise, a steepening mountain slope, a hockey-stick. Graph after graph after graph, all the same. This runaway growth is the profile of our contemporary existence. It is the universal model of the period of history that I have witnessed on Earth – the great underlying explanation of all the change that I report. My testimony is a first-person narrative of the Great Acceleration.

You look at all these graphs – this one repeating line – and you ask yourself the obvious question: how can this continue? Of course, the answer is that it can't. Microbiologists have a graph of growth that begins with the same form, and they know how it ends. When a few bacteria are placed on a bed of food in a sterile, sealed dish – a perfect environment, free from competition, sitting on abundant nutrients – they take some time to adapt themselves to the new medium – a period called the *lag phase*. This can last just one hour, or a few days, but at some point it ends suddenly – the bacteria solve the problem of how to exploit the conditions of the dish, and begin to reproduce by dividing, doubling their population as frequently as every 20 minutes. So begins the *log phase*, a period of exponential growth, the bacteria splitting and spreading in surges across the surface of the food. Each individual bacterium grabs its own plot and seizes what it needs. Ecologists call this a scramble competition – every bacterium for itself! It's a type of competition that does not end well in a closed system such as the finite, sealed dish. When the bacteria reproduce to such a degree that they reach the edge, every individual cell will begin to disadvantage every other at the same moment. The food begins to run out beneath the bacteria. Exhaust gases, heat and effluents begin to accumulate and poison with increasing speed. Cells start to die, tempering the growth rate of the population for the first time. These deaths also occur exponentially due to the worsening environment, and soon there is a moment when the death rate and the birth rate equal each other. At that point, the population has peaked, and may plateau for a period. But within a finite system, this won't continue forever – it's not *sustainable*. Food starts to run out everywhere, the gathering waste becomes deadly throughout the dish, and the colony crashes as quickly as it rose. Ultimately, the sealed dish becomes a very different place – a place with no food, its environment ruined, hot, acid and toxic.

The Great Acceleration places us, our activities and our various measures of impact in the log phase. After hundreds of millennia of lag, we humans appear to have solved the practical problems of living on Earth in the middle of the last century. It was probably an inevitable outcome of the rise of the industrial age – which enabled us, with new sources of power and machinery to multiply the efforts of an individual – but it appears to have finally been triggered by the end of the Second World War. The war effort itself was responsible for breakthroughs in medicine, engineering, science and communication. The end of the war provoked the formation of a host of multinational initiatives, including the United Nations, the World Bank and the European Union, all designed to unite the world and ensure that the global human society worked together. Such initiatives played their part in bringing an unmatched period of relative peace – the Great Peace – and it was because of this that we could exploit our freedoms, accelerating every opportunity for growth.

The Great Acceleration curve is the look of progress. During its reign, for the majority, measures of human development have risen remarkably – average life expectancy, global literacy and education, access to healthcare, human rights, per capita income, democracy. It was the Great Acceleration that brought the advances in transport and communications that made my career. The astonishing expansion in all manner of activities that we have managed to achieve in the last 70 years has brought many of the things we may have wished for. Yet we must acknowledge that, in addition to all the benefits, there are costs. Like the bacteria, we have our exhaust gases, our acids and our toxic waste. These costs also accumulate exponentially. Our accelerating growth cannot continue forever – those photographs from Apollo show quite clearly that Earth is a closed system just like the bacteria colony's sealed dish. We urgently need to know how much more our planet can take.

Some of the most important science of recent years has examined nature at a planetary scale in order to discover these details. A team of leading *Earth system* scientists led by Johan Rockström and Will Steffen has studied the resilience of ecosystems across the globe.[2] They looked carefully at the elements that have enabled each ecosystem to function so reliably during the Holocene, and tested with modelling at what point each of these ecosystems would start to fail. In effect, they have been uncovering the inner workings and inbuilt weaknesses of our life-support machine – a remarkably ambitious project that has transformed our understanding of the way the planet works.

They found nine critical thresholds hard-wired into Earth's environment – nine *planetary boundaries*. If we keep our impact within these thresholds, we occupy a safe operating space, a sustainable existence. If we push our demands to such an extent that any one of these boundaries is breached, we risk destabilising the life-support machine, permanently debilitating nature and removing its ability to maintain the safe, benign environment of the Holocene.

In the control room of Earth we are absentmindedly turning up the dials on these nine boundaries, just as the hapless nightshift crew did in Chernobyl in 1986. The nuclear reactor also had its

The Planetary Boundaries Model

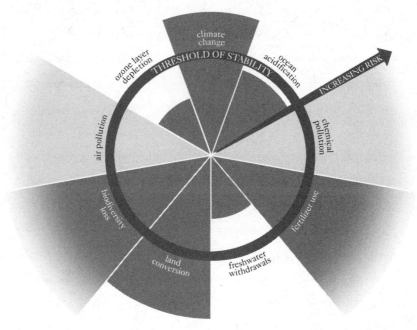

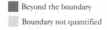

Beyond the boundary
Boundary not quantified

climate change

ozone layer depletion

THRESHOLD OF STABILITY

ocean acidification

INCREASING RISK

air pollution

chemical pollution

biodiversity loss

fertilizer use

land conversion

freshwater withdrawals

inbuilt weaknesses and thresholds, some known to the crew, some not known. They moved the dials on purpose to test the system, but did so without due respect or understanding of the risks they were taking. Once pushed too far, a threshold was breached, a chain reaction was set in motion that destabilised the machine. From that moment there was nothing they could do to stop the unfolding disaster – the complex, fragile reactor was already committed to fail.

Currently, our activities are committing the Earth to failure. We have already pushed through four of the nine boundaries. We are polluting the Earth with far too many fertilisers, disrupting the nitrogen and phosphorus cycles. We are converting natural habitats on land – such as forests, grasslands and marshlands – to farmland at too great a rate. We are warming the Earth far too quickly, adding carbon to the atmosphere faster than at any time in our planet's history. We are causing a rate of biodiversity loss that is more than 100 times the average, and only matched in the fossil record during a mass extinction event.[3]

People, quite rightly, talk a lot about climate change. But it is now clear that manmade global warming is one of a number of crises at play. The work of the Earth scientists has revealed that, today, four warning lights are flashing on the dashboard. We are already living beyond the safe operating space of Earth. The Great Acceleration, like any explosion, is about to generate fallout – an equal and opposite reaction in the living world, a 'Great Decline'.

Scientists predict that the damage that has been the defining feature of my lifetime will be eclipsed by the damage coming in the next hundred years. If we don't change course, those born today could witness the following.

2030s

After decades of aggressive deforestation and illegal burning in the Amazon basin, carried out by people who wish to secure more land for agriculture, the Amazon rainforest is on course to be reduced to 75 per cent of its original extent by the 2030s. Although still large, this may prove to be a tipping point for the Amazon, triggering a phenomenon known as *forest dieback*. The forest becomes suddenly unable to produce enough moisture from its diminished canopy to feed the rainclouds, and the most vulnerable parts of the Amazon degrade firstly into a seasonal dry forest, then an open savannah. The decline is self-feeding – the more that dieback occurs, the more it causes further dieback. The drying of the entire Amazon basin is therefore predicted to be swift and devastating.[4] The biodiversity loss will be catastrophic – the Amazon is home to one in ten of all the world's known species, meaning countless localised extinctions that would trigger domino effects throughout the ecosystem. All wild populations will be hard hit, each individual finding it more and more difficult to locate food and a mate.

Species that may have yielded drugs, new foodstuffs and industrial applications may be gone before we even know they exist. But the cost to humankind is far more profound and material. We would lose a long list of the environmental services that the Amazon has always fulfilled. Erratic flooding would become common in the basin as the tree stock dies and releases the soils it holds between its roots into the rivers. Thirty million people may need to leave the watershed, including almost three million indigenous people. The change in airborne moisture would be likely to reduce rainfall over much of South America, causing water shortages in many of its megacities, and, ironically, droughts in the farmlands created by the deforestation. Food production in Brazil, Peru, Bolivia and Paraguay would be radically affected.

The Amazon's greatest environmental service is that, for the whole of the Holocene, more than 100 billion tonnes of carbon has been locked away in its trees. The wildfires of each new dry season would release this progressively into the atmosphere. At the same time, the reduced ability of the forest to photosynthesise would mean that, each year, less carbon will be removed by the region. The additional carbon dioxide in the atmosphere will doubtless speed up the rate of global warming.

At the other end of the Earth, the Arctic Ocean is expected to have its first entirely ice-free summer in the 2030s.[5] This would result in open water at the North Pole. Even the multi-year sea ice in sheltered fjords, thick with layers of repeated freezes, may not last in the warmth and may start to disappear. The algal forests on the underside of the ice would then be cast into the water, affecting the whole Arctic food chain.

Since the Earth would have less ice, it would be less white each year, meaning less of the Sun's energy would be reflected back out to space, and the speed of global warming would increase again. The Arctic would start to lose its ability to cool the planet.

2040s

The next major tipping point is expected to occur a few years after this jump in warming has taken place. For several decades, the warming climate in the north will have been thawing the *permafrost*, the previously frozen soils that exist below the tundra and forests of much of Alaska, northern Canada and Russia.[6] It is a trend that is much harder to detect or predict than the retreat of the sea ice, yet it is potentially far more hazardous. For the entire Holocene, frozen water has constituted as much as 80 per cent of the soils in these regions. On a warmer Earth, that would not continue. The only sign of the thaw above the surface has been the appearance of new lakes and ugly craters in the far north where the land has slumped as the water has drained away. But in the 2040s there is expected to be a much wider collapse in the tundra. Within a few years, the entire north – an area that accounts for a quarter of the land surface in the northern hemisphere – could become a mud bath as the ice that held the soil together disappears. There would be massive landslides and vast floods as millions of cubic metres of newly fluid soils seek lower ground. Hundreds of rivers would change course, thousands of small lakes would be emptied. Lakes near the shore could spill into the ocean, sending giant plumes of silty freshwater out to sea. The impact on the local wildlife would be overwhelming, and the people living in the region – indigenous groups, fishing communities, oil and gas company employees, transport and forestry workers – would have to leave the area. But the key consequence of the thaw would affect everyone on Earth. For thousands of years, the permafrost has locked in an estimated 1,400 gigatonnes of carbon – four times more carbon than humankind has emitted in the last 200 years, and twice as much as there is in the atmosphere. The thaw would release this carbon, gradually, over many years, turning on a gas tap of methane and carbon dioxide that we would probably never be able to turn off.

2050s

Any wildfires and thaws occurring in the next three decades would send the carbon count of the atmosphere into its own great acceleration. As always, the surface waters of the ocean would take more than their fair share of this carbon. On entering the water, carbon dioxide forms carbonic acid, first in the shallows, then, due to the flows of ocean circulation, throughout the water column. By the 2050s, the entire ocean could be sufficiently acidic to trigger a calamitous decline.

Coral reefs, the most diverse of all marine ecosystems, are particularly vulnerable to the increasing acidification.[7] Weakened by years of bleaching events, the rising acidity will make it harder for them to repair their calcium carbonate skeletons. In an era of warmer air, and stronger storms, reefs could well be ripped apart. Some predict that 90 per cent of the coral reefs on Earth will be destroyed in the space of a few years.

The open ocean is also vulnerable to acidification. Many species in the plankton at the base of the food chain also have calcium carbonate shells. The increasingly acid ocean would inhibit their ability to bloom and flourish. Fish populations all the way up the chain would suffer as a result. Oyster and mussel harvests would start to fail. The 2050s could prove to be the beginning of the end for the remaining commercial fisheries and fish farming. The livelihoods of more than half a billion people would be directly affected, and a ready source of protein that has fed us for our entire history would start to disappear from our diets.

2080s

By the 2080s, global food production on land could be at crisis point.[8] In the cooler, wealthier parts of the world, where intensive agriculture has been adding too much fertiliser for a century, the soils would be exhausted and lifeless. Key harvests would fail. In the warmer, poorer parts of the world, global warming may

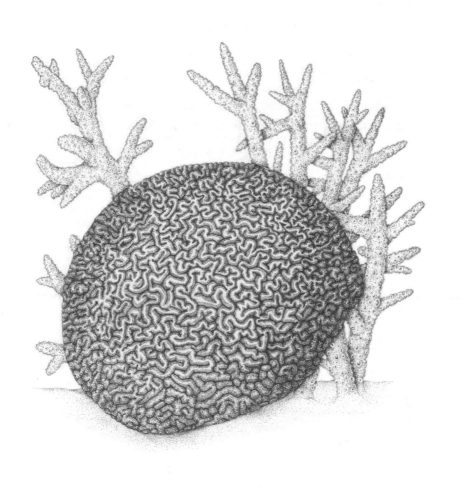

bring higher temperatures, changes in the monsoon, storms and droughts that doom farming to failure. Across the world, millions of tonnes of lost topsoil could enter the rivers and bring flooding in the towns and cities downstream.

If the current rate of pesticide use, habitat removal and the spread of diseases in pollinators like bees continues, the loss of insects species would come to affect three-quarters of our food crops by the 2080s. Nut, fruit, vegetable and oilseed harvests could fail if unable to rely on the diligent work of insects for their pollination.[9]

At some stage, the situation may well be made worse with the emergence of another pandemic. We are only just beginning to understand that there is an association between the rise of emergent viruses and the planet's demise. An estimated 1.7 million viruses of potential threat to humans hide within populations of mammals and birds.[10] The more we continue fracturing the wild with deforestation, the expansion of farmland and the activities of the illegal wildlife trade, the more likely it is that another pandemic will arise.

2100s

The twenty-second century could begin with a worldwide humanitarian crisis – the largest event of enforced human migration in history.

Coastal cities worldwide would be facing a predicted sea level rise of 0.9m during the twenty-first century, caused by the slowly

melting ice sheets of Greenland and Antarctica, together with a creeping expansion of the ocean as it warms.[11] For 50 years, over a billion people in 500 coastal cities may have already been battling storm surges, but the sea level could be high enough by 2100 to destroy ports and flood hinterlands.[12] Rotterdam, Ho Chi Minh City, Miami and many others would become impossible to defend, and hence uninsurable and uninhabitable. The evicted populations would have to move further inland.

But there is a greater problem. Should all these events unfold as described, our planet would be 4°C warmer by 2100. More than a quarter of the human population could live in places with an average temperature of over 29°C, a daily level of heat that today scorches only the Sahara.[13] Farming in these areas would be impossible, and a billion rural people may be forced to go in search of better prospects. Those parts of the world with climates that are still relatively mild would be put under excessive pressure to accept the human traffic. Inevitably, borders would be shut and conflicts would be likely to erupt globally.

In the background, the sixth mass extinction would become unstoppable.

* * *

Within the lifespan of someone born today, our species is currently predicted to take our planet through a series of one-way doors that bring irreversible change and commit us to losing the security and stability of the Holocene, our Garden of Eden. In such a future, we will bring about nothing less than the collapse of the living world, the very thing that our civilisation relies upon.

None of us want this to happen. None of us can afford to allow this to happen. But, with so many things going wrong, what do we do?

The work of scientists who study the Earth's systems gives us the answer. In fact, it's quite straightforward. It's been staring us in the face all along. Earth may be a sealed dish, but we don't live in it alone! We share it with the living world – the most remarkable life-support system imaginable, constructed over billions of years to refresh and renew food supplies, to absorb and reuse waste, to dampen damage and bring balance at the planetary scale. It is no accident that the planet's stability has wavered just as its biodiversity has declined – the two things are bound together. To restore stability to our planet, therefore, we must restore its biodiversity, the very thing we have removed. It is the only way out of this crisis that we ourselves have created. We must *rewild* the world!

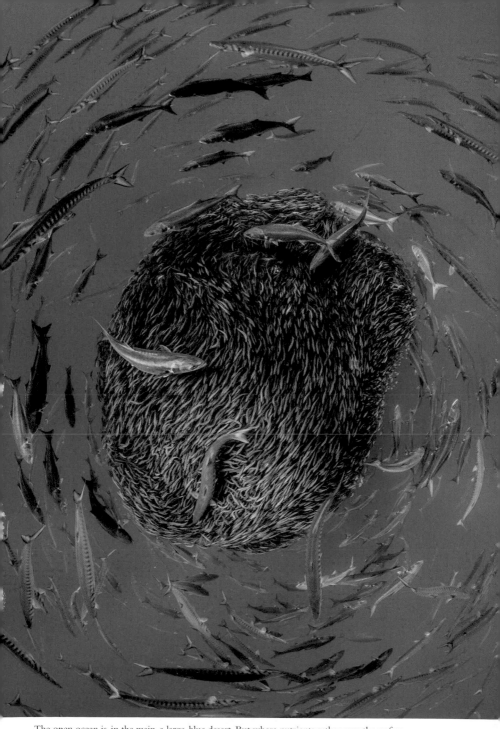

The open ocean is, in the main, a large, blue desert. But where nutrients gather near the surface, plankton bloom, leading to a flurry of activity. Here, a school of mackerel, attracted by the plankton, forms a baitball as it is pursued by barracuda and bluefish. (© Jordi Chias/naturepl.com)

Plastic ocean pollution: a Whale Shark filter feeds in polluted waters, ingesting plastic.
(© Rich Carey/Shutterstock)

A Chinese labourer sorts out plastic bottles for recycling in Dong Xiao Kou village,
on the outskirts of Beijing. (© Fred Dufour/AFP/Getty)

Plastic rubbish washed up on the beaches of Christmas Island – a remote atoll in the Pacific Ocean. (© Gary Bell/Oceanwide/naturepl.com)

A Hawaiian monk seal is caught in fishing tackle off Kure Atoll in the Pacific Ocean. The seal was subsequently freed and released by the photographer. (© Michael Pitts/naturepl.com)

The sea otter is a keystone species of kelp forests, one of the most productive marine habitats. The otters prey on sea urchins that eat the kelp, helping the seaweed forest to thrive – an example of how increased biodiversity helps natural systems to better capture and store carbon. (© Bertie Gregory/naturepl.com)

The European bison was hunted to extinction in the wild in the early twentieth century, but re-introductions from captivity are now gaining a foothold in many nations and the bison is becoming an icon of the European rewilding movement. (© Wild Wonders of Europe/Unterthiner/naturepl.com)

The coral reefs and open waters of Palau were once overfished, but strong policies based upon traditional, sustainable fishing approaches have dramatically improved marine biodiversity. (© Pascal Kobeh/naturepl.com)

A white stork landing with nesting material and joining his mate on the Knepp Estate, a pioneering wildland farm in the UK in April 2019. This is the first recorded instance of white storks nesting in the UK for several hundred years. (© Nick Upton/naturepl.com)

Dian Fossey with mountain gorillas in Rwanda. She drew the world's attention to the plight of this species of gorilla and enabled us to film them for *Life on Earth*. (© The Dian Fossey Gorilla Fund International)

Grey wolves on a ridge in Yellowstone National Park, USA. The reintroduction of wolves to the park in 1995 profoundly affected the entire ecosystem, demonstrating the value of top predators in raising the biodiversity of natural systems. (© Sumio Harada/Minden/naturepl.com)

The Ouarzazate Solar Power Station in Morocco, the world's largest concentrated solar power plant, is built to supply electricity through the night using energy stored in molten salt. (© Xinhua/Alamy Live News)

With director, and my co-author, Jonnie Hughes, in the very Leicestershire quarry that I used to visit on fossil finding expeditions when I was a boy. Here, discussing the script during the filming of the feature documentary that accompanied the release of this book. (© Ilaira Mallalieu)

I have long been a supporter of WWF. In 2016, I spoke at the launch of their Living Planet Report, the biannual health check of the Earth that has become the definitive guide to the extent of biodiversity loss on our planet. (© Stonehouse Photographic/WWF_UK)

PART THREE

A Vision for the Future

How to Rewild the World

How can we encourage a return of the wild and bring back some stability to the Earth? Those who contemplate the path to an alternative, wilder, more stable future, are unanimous in one respect: our journey must be guided by a new philosophy – or, more accurately, a return to an old philosophy. At the beginning of the Holocene, before farming was invented, a few million humans across the globe were living as hunter-gatherers, an existence that was sustainable, that worked in balance with the natural world. It was the only option our ancestors had at the time.

With the advent of farming, our options increased, and our relationship with nature changed. We came to regard the wild world as something to tame, to subdue and use. There is no doubt that this new approach to life brought us spectacular gains, but over the years, we lost our balance. We moved from being a part of nature to being apart from nature.

All these years later, we need to reverse that transition. A sustainable existence is once again our only option. But there are now billions of us. We can't possibly return to our hunter-gatherer ways. Nor would we want to. We need to discover a new kind of sustainable lifestyle, one that brings our contemporary human world back into balance with nature once again. Only then will the biodiversity loss we have caused begin to turn to biodiversity gain. Only then can the world rewild, and stability return.

We already have a compass for this journey to a sustainable future. The planetary boundaries model is designed to keep us on the right path. It tells us that we must immediately halt and preferably start to reverse climate change by attending to greenhouse gas emissions wherever they occur. We must end our overuse of fertilisers. We must halt and reverse the conversion of wild spaces to farmland, plantations and other developments. It also warns us of the other things we need to keep an eye on – the ozone layer, our use of freshwater, chemical and air pollution, ocean acidification. If we do all those things, biodiversity loss will begin to slow to a halt, and then start itself to reverse. Or to put it another way, if the chief measure by which we judge our actions is the revival of the natural world, we will find ourselves making the right decisions, and we will do so not just for the sake of nature, but, since nature keeps the Earth stable, for ourselves.

But our compass is missing an important element. A recent review has estimated that almost 50 per cent of humanity's impact on the living world is attributable to the richest 16 per cent of the human population.[1] The lifestyle that the wealthiest of us have become used to on Earth is wholly unsustainable. As we plot a path to a sustainable future, we will have to address this issue. We must

The Doughnut Model

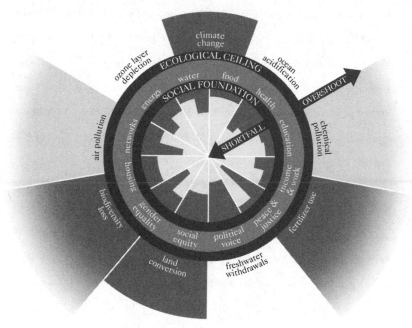

learn not only to live within the Earth's finite resources, but also how to share them more evenly too.

The University of Oxford economist Kate Raworth has clarified this challenge by adding an inner ring to the planetary boundaries model. This new ring holds the minimum requirements of human well-being: good housing, healthcare, clean water, safe food, access to energy, good education, an income, a political voice and justice. It hence becomes a compass with two sets of boundaries. The outer ring is an ecological ceiling below which we must remain if we are to have a chance of maintaining a stable and safe planet. The inner ring is a social foundation that we must aim to raise everyone above to enable a fair and just world. The resulting model has been named the *Doughnut*, and it is an enticing prospect – a safe and just future for all.[2]

'Sustainability in all things' should be our species' philosophy; the Doughnut Model, our compass for the journey. The challenge it sets us is simple, yet formidable: to improve the lives of people everywhere, while at the same time radically reducing our impact on the world. And what should be our source of inspiration in trying to meet this great challenge? We need look no further than the living world itself. All the answers are there.

Moving Beyond Growth

Our first lesson from nature concerns growth. We have arrived at this moment of desperation as a result of our desire for *perpetual growth* in the world economy. But in a finite world, nothing can increase forever. All the components of the living world – individuals, populations, even habitats – grow for a period of time, but then they mature. And once mature, they may thrive. Things can thrive without necessarily getting bigger. An individual tree, an ant colony, a coral reef community or the entire Arctic ecosystem, all exist for a prolonged period when mature as successful entities. They grow to a point, then make the most of things – exploiting their newly won positions, but in a sustainable manner. They move from the period of exponential growth, the log phase, past a peak to a plateau. And, as a result of the way they interact with the living world beyond, that stable plateau period can last indefinitely.

That is not to say that a plateauing wild community does not change. The Amazon is tens of millions of years old.[3] In that time, it has covered roughly the same patch of Earth with its vast closed canopy as it did until recently, thriving in one of the planet's prime pitches. The amount of sunlight and rainfall it has received and the level of nutrients in its soil may have been roughly constant throughout. But the species in its living community will have changed significantly in that time. Like teams shifting their position in a sports league table, or share prices on a stock exchange, in any one year there will have been winners and losers. There will always be populations on the ascent, moving into an area and multiplying at the expense of another; individual trees seizing the site where another has fallen. There will be new arrivals, and others that fade away. Some of these new arrivals may have innovations that boost the opportunities for others – a new species of bat, for example, may act as a pollinator for night-flowering plants. Conversely, the loss of species may, at the same time, reduce opportunities elsewhere in the forest. Ever adjusting, reacting and refining, the Amazon rainforest community can continually thrive over tens of millions of years without demanding any further raw resources from the Earth. It is the most biodiverse place on the planet – the most successful of life's current enterprises – but it has no need for net growth. It is mature enough to simply last.

Humankind currently appears to have no intention of reaching such a mature plateau. As any economist will explain, over the last 70 years all our social, economic and political institutions have adopted one overriding goal – an ever-increasing growth in each nation, judged by the crude measure of gross domestic product. The organisation of our societies, the hopes of business, the promises of politicians, all require GDP to climb ever upwards. The Great Acceleration is the product of this fixation, and the Great Decline of the living world, its consequence. For, on a finite planet, the only way to achieve perpetual growth is to take more from elsewhere. What felt like a miracle of the modern age was just stealing. As the appalling statistics I listed at the end of my witness statement attest, we have taken everything we have directly from the living world. And we have done this while ignoring the damage we have been doing. The species loss caused by deforestation to grow the soy we need to feed the chicken we eat is not accounted for. The impact on marine ecosystems of the plastic water bottle that we buy and discard is not accounted for. The greenhouse gases produced when making the concrete for the breezeblocks of the extension we build are not accounted for. Little wonder that all of the damage we have done to Earth has crept up on us so quickly.

A new discipline within economics is attempting to solve this problem. Environmental economists are focused on building a sustainable economy. Their ambition is to change the system so that markets around the world benefit not just profits, but also people and the planet too. They call these the three Ps. Many among them have high hopes for what they term *green growth* – a type of growth that has no negative impact on the environment. Green growth may come from making products more energy-efficient, or from turning dirty, impactful activities into clean, low or zero-impact activities, or from driving growth in the digital world, which, when powered by *renewables*, could be described as a low-impact sector.

The advocates of green growth point to a history of waves of innovation that have periodically revolutionised the possibilities for humankind. First there was the advent of water power in the eighteenth century, enabling mills to drive machinery that hugely increased the productivity of a business. Then came our adoption of fossil fuels and steam power, which not only caused an industrial revolution in manufacturing, but also brought railways and shipping and eventually aircraft that could distribute people and products quickly across the globe. Three waves followed. The electrification of the early twentieth century that brought telecommunications, the space age of the 1950s that presided over a consumer boom in the West, and the digital revolution that launched the internet and brought hundreds of smart devices into our homes. All these have radically changed the world and brought booms in business. The hope and expectation of many environmental economists is that a sixth wave of innovation – the *sustainability revolution* – is almost upon us. In this new order, innovators and entrepreneurs will make fortunes by devising products and services that reduce our impact on the planet. Of course, we are already experiencing the start of this – low-energy light bulbs, cheap solar power, plant burgers that taste like meat, sustainable investments. The hope is that, faced with the scale and urgency of our planet's Great Decline, politicians and business leaders will stop subsidising damaging industries and rapidly turn to sustainability as the popular, sensible option for growth to continue, at least for a while.

In the end, though, green growth is still growth. Will human-kind ever be able to move beyond its growth phase, mature and settle into a plateau? Can it, perhaps on the other side of that sixth wave of innovation, become like the Amazon – thriving, refining, improving sustainably over the long term, but without getting bigger? There are those who hope for a future in which humankind globally detaches itself from its addiction to growth, moves on from GDP as the be-all and end-all, and becomes focused upon a new, sustainable measure of success that involves all three Ps. The Happy Planet Index, created by the New Economics Foundation in 2006, attempts to do just that, combining a nation's *ecological footprint* with elements of human well-being, such as life expectancy, average levels of happiness and a measure of equality. When you rank countries by this index, you get a completely different league table than from GDP alone. In 2016, Costa Rica and Mexico came top, with better average well-being scores than the USA and UK at a fraction of the ecological footprint. The Happy Planet Index is certainly not foolproof. Since it is a merged score, it's possible, like Norway, to rank highly with a heavy footprint if your well-being score is very high. It is also possible, like Bangladesh, to rank highly with poor well-being, if your footprint is light. Yet the Happy Planet Index and others like it are being seriously considered by a number of nations as alternatives to GDP, and encouraging a wider debate about the sum purpose of all humankind's efforts on Earth.[4]

In 2019, New Zealand made the bold step of formally dropping GDP as its primary measure of economic success. It didn't adopt any of the existing alternatives, but instead created its own index based upon its most pressing national concerns. All three Ps – profit, people and planet – were represented. In this single act, Prime Minister Jacinda Ardern shifted the priorities of her whole country away from pure growth and towards something that better reflects the issues and aspirations many of us have today. The change in agenda may have made her decisions more straightforward when coronavirus arrived in February 2020. She locked the country down before there had been a single death, while other nations hesitated, nervous perhaps, of the effects on the economy. By early summer, New Zealand had few new cases, and could go back to work and mix freely.

New Zealand may be a guiding light. Surveys in other nations show that people across the world are now keen for their governments to prioritise people and planet over profit alone. It is an indication that voters and consumers everywhere may be ready for a sustainable, and ultimately, as Kate Raworth terms it, growth-agnostic world. Every nation has a journey to make to become prosperous and good for its people and good for the planet. The wealthy nations that have benefitted from unsustainable growth have the formidable task of maintaining a good standard of living whilst radically reducing their footprints. Poorer nations have the very different challenge of radically raising their standards of living in a way that's never been done before – whilst achieving a sustainable footprint. Through this lens all nations are now developing nations with work to do, and all will need to switch to green growth, and join the sustainable revolution.

Humankind has yet to mature. Like a sapling in the Amazon eagerly grasping its opportunity to take over a clearing, we have concentrated all our efforts to date on growth. But, according to the environmental economists, we must now curb our passion for growth, distribute resources more evenly and start to prepare for life as a mature canopy tree. Only then will we be able to bask in the sunlight that our speedy development won for us, and enjoy an enduring, meaningful life.

Switching to
Clean Energy

The living world is essentially solar-powered. The Earth's plants, together with phytoplankton and algae, capture three trillion kilowatt hours of solar energy every day. That is almost 20 times the energy we use. And they collect it directly from sunlight, trapping the energy within organic molecules made from carbon. They obtain this carbon by absorbing carbon dioxide from the air. As they build the organic molecules, they expel oxygen as a waste product. The process is known as photosynthesis. It powers all their life processes, from the growth of their stems and trunks, to the production of seeds to establish the next generation, fruits to persuade animals to transport their seeds, and larders in which to store their food to sustain themselves during hard times.

Animals, including ourselves, spend a lot of time trying to collect a share of this industry. We bite into the fruits that some plants produce and suck out the sugar or nibble the softer parts of their leaves and roots. We and many other animals also eat the flesh of those that feed on the plants and so collect the Sun's energy second-hand. There are even some organisms – the fungi and bacteria – that live by slowly liquefying the bodies of dead animals to collect the precious organic molecules they contain. And when any of us – animal, plant, alga, phytoplankton, fungus, or bacterium – finally come to break up these organic molecules to get at the energy within, carbon dioxide escapes as a by-product into the atmosphere to be used by plants in photosynthesis once again.

The capture and distribution of the Sun's energy, and the cycling of carbon between the atmosphere and the living world that results, has been central to the activity of life on Earth for 3.5 billion years. In that time a host of forests, marshes, swamps, mats and blooms have brought power to the living world of their day. As they died, the carbon that they contained was returned to the atmosphere through the process of decomposition. But there have been times when this cycle has been disrupted and decomposition failed to occur. The first plants large enough to be described as trees appeared on Earth around 300 million years ago. They resembled the tree ferns and horsetails that are their relatively tiny living descendants. These first forests grew in tropical freshwater swamps that covered much of the planet's land. As the trees died, their bodies fell into the swamps and accumulated underwater, being slowly entombed by sediment brought down by the rivers. Beyond the reach of oxygen and the normal processes of decomposition, their carbon-laden tissues, buried beneath mud and sand, were compressed and eventually became coal. Subsequently, over several hundred million years, plankton and algae that flourished in ancient seas and stagnant lakes have, on occasions, been buried at depth and turned into oil and inflammable gas.

Two hundred years ago, we started to dig up these energy-rich remains and burn them, returning great quantities of the carbon they contain to the atmosphere as carbon dioxide. We have learned to harness this fossil fuel energy so skilfully that, today, our homes are heated by it, our vehicles driven by it, and our factories are powered by it to such an extent that we can melt steel should we wish to do so. The sunlight of those billons of long-past days has fuelled our Great Acceleration. But in the process we have returned millions of years-worth of carbon back into the atmosphere in a matter of decades.

It is a potentially disastrous thing to have done. Carbon dioxide in itself is a relatively inactive, innocuous gas. We breathe it out with every breath. But it is a greenhouse gas – that is to say, it acts in the atmosphere like a blanket, trapping heat close to the Earth's surface. The greater its concentration, the more effective it is at warming the Earth. Carbon dioxide also dissolves in water, and has the effect of increasing the acidity of the ocean. By overloading the carbon in the atmosphere, we are, in fact, replicating the changes that led to the greatest ever mass extinction, at the end of the Permian. However, we are bringing about these changes at a much faster rate.

We suddenly find ourselves at a huge disadvantage. We now have no option but to change the way in which we power our activities. Yet there is little time for us to do so. In 2019, fossil fuels provided 85 per cent of our global energy.[5] Hydropower, which is low-carbon but limited to certain locations and capable of doing significant environmental damage, provided under 7 per cent. Nuclear power, which is also low-carbon, but certainly not without its risks, provided just over 4 per cent. The power sources that we should be using, the inexhaustible natural sources of energy – the Sun, the wind, the waves, the tides and the heat from deep in the Earth's crust – the so called renewables – are still used for just 4 per cent of our capacity at present. We have less than a decade to switch from fossil fuels to clean energy. We have already increased global temperature by 1°C from pre-industrial levels. If we are to halt its increase at 1.5°C, there is a limit to the amount of carbon we can yet add to the atmosphere – our *carbon budget* – and, at current emissions rates, we will add this amount before the end of the decade.[6]

Our careless use of fossil fuels has set us the greatest and most urgent challenge we have ever faced. If we do make the transition to renewables at the lightning speed required, humankind will forever look back on this generation with gratitude, for we are indeed the first to truly understand the problem – and the last with a chance to do anything about it. The road to a world powered by carbon-free energy will be a bumpy one, and the next few decades will be extremely challenging for us all. But many working on this problem believe it is possible. We human beings are, above all, the most astonishing problem solvers. We have made difficult journeys before that evolve enormous social change throughout our history, and we can do so again.

The first barrier to progress is already largely overcome – that of a practical alternative. The energy sector now has a good understanding of how to generate electricity from the Sun, wind, water and the natural heat of the deep Earth. Outstanding issues remain. There is still a storage problem. Battery technologies are not yet adequately developed. Nor are renewables as efficient as they need to be to provide fully for the tasks of transport, heating and cooling. On such occasions we have to bridge our shortcomings with temporary fixes that will help us get around the problem. Sometimes, these bridges come with what Paul Hawken of Project Drawdown[7] describes as 'regrets'. It is likely that we will bridge our current shortcomings with nuclear power, large hydropower and a prolonged use of natural gas, which is a fossil fuel, but produces far fewer carbon-rich emissions than coal or oil. All come with some regret. We may develop *bioenergy* solutions, in which agricultural products are used as an energy source, but that too comes with some regret, because its production would require huge swathes of land. For transport fuel, it may be that hydrogen fuel cells and sustainable biofuels made from plant and algal oils will join electric vehicles to become a permanent part of the mix for road, rail and shipping. Most experts agree that air transport will be the most difficult problem to solve. Hybrid, fully electric and hydrogen planes are in development, but until they are viable at scale, airlines are planning to build *offsets* of carbon emissions into ticket prices. We have to work hard to ensure that all these fixes are as temporary as possible. With so little time before we use up our carbon budget, any continued use of fossil fuels inevitably demands sharper, deeper cuts in emissions elsewhere.

A second potential barrier is affordability, but this too is falling away. The scaling up of solar and wind power has already brought the price of renewable generation per kilowatt down to levels that outcompete coal, hydropower and nuclear, and it is approaching the cost of gas and oil. In addition, renewables are much cheaper to manage than other power sources. Over 30 years, it is estimated that a renewable-dominated energy sector would save trillions of dollars in operational costs. Many commentators believe that improving affordability alone will mean that renewables will swiftly replace fossil fuels. But there is a third barrier that they may have been underestimating

Perhaps the most formidable obstacle we face is the abstract force we might call vested interests. Change is a threat to any invested in the status quo. Currently, six of the ten largest companies in the world are oil and gas companies. Three of these are state-owned, and two of the other four are concerned with transport. But they are far from the only ones reliant on fossil fuels. Almost every large company and government uses fossil fuels predominantly for their power and distribution. Most heavy industry uses fossil fuels for heat or to cool products in its production lines. Most of the large banks and pension funds have invested heavily in fossil fuels, the very things that are jeopardising the future we are saving for. To bring about change in a system as entrenched as ours is will take a number of carefully judged steps. Those who analyse energy transition predict that banks, pension funds and governments will increasingly release their coal and oil stock, in an attempt to avoid huge losses. Politicians will be called upon to divert the hundreds of billions of dollars in subsidies that currently go to the fossil fuel sector, to help push for renewables. Local governments have already started to pay attractive rates to households that generate their own electricity for any surplus and to assist communities in creating their own renewable *micro-grids*.

Other trends that are hard to spot from today's vantage point may also prove to be highly significant in speeding up the move away from fossil fuels. Some analysts predict that the advent of autonomous vehicles will revolutionise the transport sector.[8] Within only a few years, they expect city dwellers to abandon car ownership and order a car only when needed. These cars would all be electric, they would charge themselves from clean energy and might be managed directly by the car manufacturers themselves, encouraging the entire industry to improve its efficiency and reliability.

It is widely acknowledged that the most powerful incentive of all to end our reliance on fossil fuels would be a high global price on carbon emissions – a *carbon tax* that penalises any and all emitters. The Swedish government introduced such a tax in the 1990s, and it led to a strong move away from fossil fuels in many sectors. The Stockholm Resilience Centre[9] suggests that a rising price, starting at $50 per tonne of carbon dioxide emitted, would be enough to stimulate rapid change from dirty to clean technology, trigger efficiency drives in those practices still dependent on fossil fuels, and excite the sharpest minds to search for new technologies and practices that lower emissions. We should be careful to do this in a way that protects the poorest in society, but studies show that is entirely achievable.[10] A carbon tax would, in short, radically speed up the sustainable revolution we need.

As the new, clean, carbon-free world comes online, people everywhere will start to feel the benefits of a society run on renewables. Life will be less noisy. Our air and water will be cleaner. We will start to wonder why we put up for so long with millions of premature deaths each year from poor air quality. Poorer nations that still have forests and grasslands could benefit from selling their carbon credits to those still dependent upon fossil fuels. They could then build renewables and low-emission life into the design of their development. Perhaps one day their smart, clean cities may become some of the best places on Earth to live, attracting the brightest stars of each generation.

Is this fantasy? It doesn't have to be. At least three nations – Iceland, Albania and Paraguay – already generate all their electricity without using fossil fuels. A further eight nations use coal, oil and gas for less than 10 per cent of their electricity. Of these nations, five are African and three are from Latin America. The energy transition, and the sustainable revolution in general, offer rapidly developing countries an extraordinary opportunity to do things differently and leapfrog many in the Western world.

Morocco is an example of a nation embracing the revolution. At the turn of the century, it relied on imported oil and gas for

almost all of its energy. Today, it generates 40 per cent of its needs at home from a network of renewable power plants, including the world's largest solar farm. It is leading the way in a promising and relatively inexpensive type of energy storage, molten salt technology, which uses plain salt to hold solar heat for many hours, enabling solar power to be used through the night. Sitting on the edge of the Sahara, and with a cable linking directly into southern Europe, Morocco could perhaps one day be a net exporter of solar energy. For a nation that was never blessed with fossil fuels, it is a ticket to a more prosperous world.

History shows that, with the right motivations, profound change can happen in a short period of time. There are signs that it is starting to happen with fossil fuels. Globally, we passed the peak of coal use in 2013. The coal industry is now in crisis as investors pull out of the sector. *Peak oil* is predicted to come in the next few years, and the plummeting prices associated with the coronavirus outbreak may even bring it closer. We may yet pull off a miracle and move to a clean energy world by the middle of this century.

There is one additional reason for hope in this respect – the possibility that, as a planet-saving bridge while we roll out clean energy, we can actively grab some of the carbon we have released into the air and lock it back out of harm's way. This *carbon capture and storage*, or CCS, is extremely attractive to politicians and business leaders who need to buy more time to phase out fossil fuels. There are filters that trap some of the carbon as it flows from fossil fuel power stations, towers of fans that remove it straight from the air, bioenergy power plants that recover greenhouse gases as the crop is digested, and facilities that pump the carbon down into rock at depth out of harm's way. Some *geoengineers* suggest more experimental ideas involving harnessing blooms of bacteria and algae, fertilising the ocean with iron, pumping CO_2 down to the bottom of the sea, and blocking the Sun with dust in the upper atmosphere. Some of these may theoretically work, and a few might be able to work at scale, but so far they are very poorly understood and risk coming with as yet unforeseen negative consequences.

What is clear to those of us concerned not only with climate change but also biodiversity loss is that we have a much better way of capturing carbon: the rewilding of the world will suck enormous amounts of carbon from the air and lock it away in the expanding wilderness. When executed in parallel with global cuts in emissions, this *nature-based solution* would be the ultimate win-win – *carbon storage* and biodiversity gain all in one. Studies in many habitats have shown that the more biodiverse an ecosystem, the better it captures and stores carbon.[11] Nature-based carbon capture is where governments, fund-managers and businesses should be investing. This is where all our offsets should go – a globally funded and internationally supported drive to revive the wild world. It would work vigorously in every habitat on Earth, halting climate change and the sixth mass extinction at the same time. Some of the swiftest gains could be won within only a few years, most spectacularly within the greatest wilderness of them all.

Rewilding the Seas

The ocean covers two-thirds of the surface of the planet. Its great depths mean that it contains an even greater proportion of inhabitable space. So there is a special role for the ocean in our revolution to rewild the world. By helping the marine world to recover, we can do three things we desperately need to do simultaneously – capture carbon, raise biodiversity and supply ourselves with more food. It starts by working with the industry that is currently causing most damage to the ocean – fishing.

Fishing is the world's greatest wild harvest, which means that, if we do it right, it can continue, because there is a mutual interest at play – the healthier and more biodiverse the marine habitat, the more fish there will be, and the more there will be to eat. So why isn't it working at present? We fish some places and some species too much. We waste too much. We use clumsy fishing techniques that wreck the ecosystem. And most damaging of all, we fish everywhere. There is nowhere in the ocean left to hide. Marine biologists such as Professor Callum Roberts explain that all of these issues can be fixed if we adopt a global approach directed by the information we already have from marine science.

Firstly, we should create a network of no-fish zones throughout coastal waters. At present there are over 17,000 *Marine Protected Areas*, or MPAs, around the world. But these only account for less than 7 per cent of the ocean, and in many MPAs certain types of fishing are still permitted.[12] It's imperative that a healthy proportion of the ocean is not fished at all, due to the way that fish reproduce. No-fish zones allow individual fish to grow older and bigger. And bigger individuals produce disproportionately more offspring. They then, in turn, repopulate neighbouring waters that are fished. This *spill-over effect* has been shown to occur around strict MPAs from the tropics to the Arctic. Fishing communities tend to resist fishing restrictions when they are first put in place, but, within a few years, they will start to feel the benefits.

The Marine Protected Area of Cabo Pulmo lies at the tip of Baja California in Mexico. In the 1990s, this area of sea was extensively overfished, and the fishing community, desperate for a solution, agreed to suggestions from marine scientists to set aside over 7,000 hectares of their coast as a no-fish zone. The local people describe the years immediately after the MPA was opened in 1995 as the hardest years they had ever faced. The fishing families caught very little in the neighbouring waters and had to survive on food vouchers offered by the Mexican government. Fishermen could see growing shoals in the MPA, and were often tempted to break the ban. It was only the faith the community had in the marine scientists that kept their resolve. It was at about the ten-year point that sharks came back to Cabo Pulmo. The older fishermen remembered them from their childhood, and knew they were a sign of recovery. After only 15 years, the amount of marine life in the no-fish zone had increased by more than 400 per cent to a level similar to reefs that had never been fished at all, and the fish shoals began to spread into the neighbouring waters. The fishermen caught more fish than they had done in decades, and what is more, the community had a tourist attraction on their doorstep. The men and women of Cabo Pulmo found new sources of income – dive shops, guesthouses and restaurants.[13]

The MPA model works because it stops us doing something we should never have begun to do – eat into the core fish stocks, the capital of the ocean. When there are no-fish zones within a legal fishing area, we become limited to living off the interest only. Any financier would tell you that that is a sensible, sustainable approach. And since no-fish zones increase the abundance of all fish populations, the capital gets bigger and bigger, leading to more and more interest – more fish for the net. It becomes easier to catch fish and that reduces the amount of fossil fuels expended out at sea, less unwanted by-catch, and the freedom to stay onshore when the seas look rough. Well-designed and effectively managed MPAs are a ticket to a new, healthy fishing relationship with the ocean. Estimates suggest that no-fish zones encompassing a third of our ocean would be sufficient to enable fish stocks to recover and supply us with fish for the long term.

The best locations for these MPAs are the places in which marine animals find it easiest to breed, the nurseries of the ocean: rocky and coral reefs, submarine seamounts, kelp forests, mangroves, seagrass meadows and saltmarshes. We should leave the waters around such places to thrive and fish the seas that neighbour them. It is no coincidence that these are also the best places to help us achieve our other big objective – carbon capture. At present, in their depleted state, saltmarshes, mangroves and seagrass meadows alone remove the equivalent of around half of all our transport emissions from the air.[14] Protected in no-fish zones, these habitats will recover to capture even more.

The way we catch fish is also important. At present, much of our fishing is far too indiscriminate. We need smarter fishing in which trawl nets have emergency exits for non-target species, in which large, predatory fish such as tuna are pole and line caught, and in which the destructive dredging of the seabed is banned. We need to constantly monitor our key fish stocks and have the self-restraint to keep within sustainable yields.[15] We should encourage new *blockchain* methods of tracking fish from dock to dish so that we can be sure where our fish comes from and choose to reward businesses that fish sustainably.

Ultimately, the aim should be to fish forever, not just to turn a quick profit – to respect the fact that wild-caught seafood is a shared resource from which all of us should be able to benefit, especially the 1 billion people, mostly from poorer communities, who rely on fish as their primary source of protein. This

ambition of taking what you need, rather than what you can get, runs through the traditions of the people of Palau, an island nation in the tropical Pacific. Having lived on their archipelago for 4,000 years, separated from the rest of the world by hundreds of miles of deep water, the sustainability of their fish stocks has always been their ultimate concern. For generations the elders have carefully monitored the fishing on their reefs and acted quickly should any one stock start to decline. They use the ancient rule of 'bul', or prohibition, to turn a reef into a no-fish zone overnight, and refuse to lift it until the neighbouring waters are busy with the reef's fish once more.

This tradition now sits at the heart of the country's fishing policies. Their four-time president, Tommy Remengesau Jr., describes himself as a fisherman taking a leave of absence to serve in government. He has seen the population of his nation boom, the tourists start to arrive, and the commercial fishing fleets from Japan, the Philippines and Indonesia wander into Palau's waters. When the demand on the ocean got too great, he did what any elder in Palau would do – he closed down the fishing. Fishing was banned entirely on some reefs and limited to low-impact practices on others, while seasonal bans were created to enable threatened fish to breed in peace. But it was what Remengesau decided for the deep waters of Palau that was most impressive. He announced that Palau should not feel obliged to keep exporting fish. It should instead plan to take just what it needed for its people and its visitors to eat – in other words, return to subsistence fishing. He radically reduced the number of

commercial licences available and turned four-fifths of Palau's terri-
torial waters, an area the size of France, into a no-fish zone. A small
number of boats continue to catch just enough tuna in the remaining
fifth for all the Palauans and their tourists. Remengesau is proud
that, due to the spill-over effect, the Palauans are offering a gift of
ever-renewing fish stocks to their neighbours.

There is now a huge opportunity for such wisdom to preside
over two-thirds of the ocean – an area that constitutes half of Earth's
surface. International waters – the high seas – are owned by no one.
They are a shared space in which all states are free to fish as much
as they wish. And that is the problem. A few nations have become
committed to paying billions of dollars in subsidies to their fleets
on the high seas. These subsidies keep the boats fishing, even when
there are too few fish left for the work to be profitable. In effect,
public money is being used to empty the open ocean. The worst
offenders are China, the EU, the USA, South Korea and Japan, all
nations that can afford to end this practice. And that is the hope –
as I write, the United Nations and the World Trade Organization
are working on a new set of rules for the high seas.[16] They are
committed to bringing an end to harmful fishing subsidies and to
delivering some respite to the overfished populations that reside in
the deep waters of the world. But it is quite clear that we could go
much further. If all international waters were designated a no-fish
zone, we would transform the open ocean from a place exhausted
by our relentless pursuit to a flourishing wilderness that would
seed our coastal waters with more fish and help us all, through its

diversity, in our efforts to capture carbon. The high seas would become the world's greatest wildlife reserve, and a place owned by no one would become a place cared for by everyone.

But we are already past the point at which this kind of approach alone is appropriate: 90 per cent of fish populations are either overfished or fished to capacity. This can be seen in the records of global catch over the past few years. We reached another peak – *peak catch* – in the mid 1990s, even as *The Blue Planet* was being filmed. From that time, we have been unable to take more than around 84 million tonnes of fish from the ocean. Yet, of course, demand for fish has continued to rise as the world population and average income has grown. Where have we been getting our extra fish? From the mid 1990s, the practice of fish farming, or *aquaculture*, has grown exponentially. In 1995, it provided 11 million tonnes of seafood. Today, aquaculture in total provides 82 million tonnes of food.[17] We have effectively doubled our catch with fish farming.

Potentially, we could use aquaculture to reduce wild seafood demand globally where we need to, but our industrial approach to date has been rife with unsustainable practices. Coastal habitats such as mangroves and seagrass meadows have been removed to make way for fish farms that hug the shore. The crop – primarily fish, prawns and clams – is frequently densely packed and diseases have been commonplace, forcing farmers to use antibiotics and disinfectant, which then can spread, with the disease itself, into the surrounding seawater. Predatory fish such as salmon have been fed on hundreds of thousands of tonnes of baitfish which we remove from the ocean, denying food to wild fish populations, a practice as bad for the ocean as overfishing. The farms can produce vast quantities of effluents which drop out of the pens and into the surrounding water. In 2007, China's vast shrimp fishery alone created 43 billion tonnes of effluents, over-fertilising the shallow seas, creating algal blooms that drain the coasted waters of their oxygen. Some farms are awash with the toxins carried by rivers, and food poisoning scares have been known. Non-native species frequently escape from farms, causing havoc among the fragile ecosystems of foreign waters.

Best practice in today's marine aquaculture sector is, to its credit, responding to all these issues.[18] Such producers show us how we may soon farm seafood sustainably. Their fish pens are spread out in the sea to dilute their impact, many located miles offshore to benefit from stronger currents. The fish within are raised in much lower densities to reduce disease, and vaccinated so that anti-biotics don't enter the water. Predatory fish are fed on oils from agricultural crops and insect protein from *urban farms* that raise billions of flies on the food waste of coastal cities. The fish farms are multi-layered, with cages of sea cucumbers and urchins – both popular foods in Asia – hanging below the fish pens, living on their falling waste. Surrounding the pens there are ropes covered with mussels and clams and hanging fronds of edible seaweeds, all benefitting from any excess food and waste carried away from the pens on surface currents.

The potential for local communities along the world's coasts to turn to these sustainable methods to increase the food and income they can get from the sea without damaging the local environment is breathtaking. There may well be ocean farmers setting up just offshore along your nearest coastlines in the near future.

And perhaps they will be joined by *ocean foresters* too. Kelp is the fastest growing seaweed on Earth, able to increase the length of its broad, brown fronds by as much as half a metre in a single day. It thrives in cool, nutrient-rich coastal waters, forming vast submerged forests that boast remarkable levels of biodiversity. Swimming through one of these forests, pushing aside the towering, leathery fronds, is an extraordinary experience. You never know quite what will be revealed as the kelp wipes across your mask! The forests are prone to attacks from sea urchins and, in cases where we have eliminated animals such as sea otters that eat the urchins, entire kelp forests have been devoured by them. But with our help, we could restore these forests and benefit significantly as a consequence. As it grows upwards, the kelp would become a home to invertebrate and fish populations, and crucially capture enormous amounts of carbon. Experiments show that each dry tonne of kelp contains the equivalent of one tonne of carbon dioxide. We could sustainably harvest the kelp as it grows and use it as a new source of bioenergy. Unlike bioenergy crops on land, the returning kelp forests would not be competing with us or terrestrial wilderness for space. When combined with CCS technology that captures carbon dioxide as the kelp is digested, we begin to enter a new territory. At that point, our power generation can actually remove carbon from the atmosphere.[19] Alternatively, the kelp could also be harvested as food for humans, feed for livestock or fish, or in order to extract its useful biochemicals. The viability of ocean forestry on a big scale is currently being investigated by a number of research groups, so we shall soon discover whether this is a possibility. What is certainly true is that if we stop overexploiting the ocean, and begin to harvest it in a way that allows it to thrive, it will help us to restore biodiversity and stabilise the planet at a speed and scale we could not hope to achieve on our own. Better-managed fisheries, a well-designed network of MPAs, support for local communities that wish to sustainably manage their coastal waters and the restoration of mangroves, seagrass meadows, saltmarshes and kelp forests around the world are the keys to achieving this.

Taking up Less Space

The conversion of wild habitat to farmland as humankind expanded its territory throughout the Holocene has been the single greatest direct cause of biodiversity loss during our time on Earth. The vast majority of this conversion has occurred in recent times. In 1700, we farmed only about 1 billion hectares of the land surface. Today, our farmland covers just under 5 billion hectares, an area equivalent to North America, South America and Australia combined.[20] This means that we currently reserve over half of all the habitable land on the planet just for ourselves. In order to gain the extra 4 billion hectares in the last three centuries, we have torn down seasonal forests, rainforests, woodland and scrub, drained wetlands and fenced in grasslands. This habitat destruction has not only been the lead cause of biodiversity loss, it has been, and continues to be, one of the lead causes of greenhouse gas emissions. The world's land plants and soils combined contain two to three times as much carbon as the atmosphere.[21] By tearing down trees, burning forests, dredging wetlands and ploughing wild grasslands, we have released two-thirds of this historic stored carbon to date. Removing the wild has cost us dearly.

Even once established, modern, industrial farmland is no substitute for wild land. It's easy to look over farmland and think of it as a natural landscape, but it is in fact very unnatural. Farmland and wild habitats function in completely different ways. Wild habitats have evolved to sustain themselves. Plants in an ecosystem cooperate to capture and store all the precious ingredients of life – water, carbon, nitrogen, phosphorus, potassium and others. Such communities have to be self-sufficient and build for the future. Over time they lock away carbon, become more complex in structure, more biodiverse, and their soils become rich in organic material.

Modern, industrial farmland is very different. We sustain it. We give it everything we think it needs and take away everything it doesn't. If the soil is poor, we add fertilisers, sometimes to the extent that it actually becomes toxic to soil micro-organisms. If there isn't enough water, we bring it in from elsewhere, reducing the water in natural systems. If other plants grow on the site, we kill them with herbicides. If insects are slowing our crop's growth, we remove them with pesticides. At the end of the growing season, we frequently strip off all the plants, and turn over the soil, exposing it to the air and sun, depleting its carbon stock. We leave herds of animals on pasture for years until the grasses have lost all their reserves and are exhausted. Farmland is supplemented territory. There is no inherent need for it to build for the future. Over time most industrially farmed land will emit carbon, become simpler in structure, lose its soil biodiversity and its organic material.[22]

Much as we may find them attractive, rolling hills of open fields, vineyards and orchards are sterile environments compared with the wilderness they have replaced. The truth is that we can't hope to end biodiversity loss and operate sustainably on Earth until we cease the expansion of our industrial farmland. Indeed, if we are to allow nature to begin to recover, we have to go further and actively reduce the proportion of the land surface that we occupy, so that we may give space back to the wild. How can we hope to do this? We all need to eat, and as the populations grow and standards of living improve, the amount of food we need will only increase. As we shall see later, addressing the immense amount of food we waste will certainly help, but even so, food industry experts have calculated that we will need to produce more food in the next four decades than all the farmers in history harvested over the entire Holocene. There is a critical question to answer: how can we get more food from less land?

There are some inspiring farmers in the Netherlands who are amongst the best-placed people to tell us. The Netherlands is one of the world's most densely populated countries. Its modest land surface is covered with farms that are smaller than in many industrialised countries with no room for expansion. In response, Dutch farmers have become expert at getting the most out of every hectare. This has come with great environmental cost but some of these farming families tell a story of change over the last 80 years that could provide inspiration for agriculture across the globe.

In the 1950s, as a result of the traumas of the Second World War, there was a strong desire in the Netherlands for families to be self-sufficient and have enough land to grow their own food. Their modest farms typically had a few animals, some cereals and some vegetables. When the next generation inherited the farms in the 1970s, they industrialised by turning to products that were increasingly available at the time: fertilisers, greenhouses, machinery, pesticides and herbicides. Each farm came to specialise in one or two crops, and families became very good at maximising yield. But their productivity relied upon diesel and chemicals. This, so far, is a similar tale to farming around the world. Biodiversity, water quality and other environmental measures worsened greatly. But then, around the millennium, their children took over and some pioneers within this generation had a new ambition – to continue increasing yield, while reducing the impact on the environment. The new young owners erected wind turbines or dug geothermal wells down below their farms to heat their greenhouses with renewable energy. They put in automated climate-control systems to keep the greenhouses at the perfect temperature while reducing water and heat loss. They began to collect all the rainwater they needed from their own greenhouse roofs. They planted their crops not in soil, but in gutters filled with nutrient-rich water to minimise input and loss. They exchanged pesticides for measured releases of natural predators, so that home-grown bee colonies

could pollinate the crops safely. When operating out in open fields, they started to measure the water and nutrient content of every square metre so that they could help keep the soils as hydrated and healthy as possible. They learnt how to make their own fertiliser and even packaging for their crop from the stems and dead leaves left over after harvest.

These innovative, sustainable farms are now among the highest yielding and lowest impact food producers on Earth. If all farmers in the Netherlands and indeed the rest of the world farmed with the ethos of these pioneering Dutch families, we would be able to produce much more food on much less land.[23] However, the high-tech approach is expensive to put in place. While it may prove to be inspirational for the large food production companies that preside over much of the world's farmland, it won't help smaller-scale and subsistence farmers. For these farmers, there are effective, low-tech approaches proven to improve yield and reduce impact in different situations around the world. *Regenerative farming* is an inexpensive approach able to revive the exhausted soils of most fields by bringing organic matter rich in carbon back into the topsoil.[24] Regenerative farmers don't till or plough because that exposes the topsoil and releases carbon into the atmosphere. They phase out the use of fertilisers since these tend to reduce soil bio-diversity and prevent the soil from functioning healthily. They sow diverse 'cover crops' after harvest to shelter the soil from direct sunlight and rainfall, and to channel nutrients through the roots of the plants back into the ground. They rotate crops in any one field over the years, using a cycle of up to ten different species of crop plant, each demanding a different profile of nutrients from the soil, so that it will never become exhausted. Crop rotation also reduces pest infestations, so that the use of pesticides can be reduced. The farmers may even intercrop, placing alternating lines of more than one crop in the same field, which together feed the soil rather than depleting it. These techniques will eventually revive depleted soil, remove the need for fertilisers altogether, and capture carbon from the air and return it to the ground. There are approximately half a billion hectares of fields worldwide that have been abandoned due to exhaustion, mostly in the poorer nations of the world. Regenerative agriculture could help them to become productive land once again, while locking away an estimated 20 billion tonnes of carbon.

Beyond the fields, there is a wave of farmers now producing food from spaces that we already occupy for other purposes. Urban farming is the practice of growing food commercially within cities. Urban farmers now grow food on rooftops, in abandoned buildings, underground, on office window ledges, down the exterior walls of city buildings, in shipping containers on brownfield sites and even above car parks, providing shade to the cars below. The farms tend to use climate-control, low-energy lighting and *hydroponics* to maximise growing conditions and keep the need to add soil, water and nutrients to a minimum. As well as making good use of wasted space, urban farms are located in the same place as their customers, so transportation emissions are greatly reduced.

A large-scale development of this approach is *vertical farming*, in which layers of different plants, often salad crops, are placed one on top of the other, lit with LEDs powered by renewables, and supplied with a nutrient medium via feeder pipes. Setting up vertical farms is expensive, but they have advantages. They multiply the yield of a hectare by up to 20 times. They do not suffer from changes in the weather and they can be sealed environments, kept free of herbicides and pesticides. Several commercial operations are already running, supplying low-volume, high-value foodstuffs like salad leaves to customers in the surrounding cities.

* * *

With the gains we can make from all these farming innovations, we can certainly boost crop yields worldwide, whilst lowering emissions. But the truth is that these improvements, even when combined with measures to limit food waste, will only get us so far. If 9 to 11 billion people are to live sustainably on Earth, there will have to be a change in the food we eat. What we eat will become more important than how much we eat. Once again, nature can explain.

On the great plains of Africa, herds of Thomson's gazelles spend much of their days eating grass. To do so, they have to expend energy locating the best shoots, biting off and chewing through the tough, outer edges of the blades to get at the sustenance inside. They only eat the blades above the ground, missing the root stock and growing point below the soil. They lose further energy as heat as they digest the grass in their stomachs, and much of the fibre in the grass passes largely undigested through their bodies and is expelled as faeces. Like all plant-eaters, the gazelles are only able to use a proportion of the energy that the plants they eat have captured from the Sun. There is an inefficiency, a loss of energy between the plants and the herbivores. Which explains why cows and antelopes have to spend much of their days eating.

A loss of energy between levels on the food chain also happens between herbivore and carnivore. Cheetah are the only predators fast enough to catch a Thomson's gazelle in full flight. They spend much of their day looking for opportunities to do so. Even when they begin a chase, they will fail to catch their prey in most instances. And when they do succeed, they will only be able to benefit from a small proportion of the energy that the gazelle has absorbed from the grass. Most of that energy will have already been spent by the gazelle, moving around looking for grass, inter-acting with other members of the herd, and indeed looking out for and evading cheetah. Furthermore, the cheetah would normally only eat the gazelle's flesh, and therefore miss all the stored energy in its bones, sinew, skin and hair.

This loss in energy as we rise up the food chain explains the numbers of animals we find in the wild. For every single predator on the Serengeti there are more than 100 prey animals. The realities of nature mean that it isn't possible for large carnivores to be common.

We humans are neither herbivores nor carnivores. We are omnivores, anatomically equipped to digest both animals and

plants. But as people become wealthy around the world, there is a trend for the size and balance of their diet to shift. Such people eat more meat each year, and this is at the heart of our unsustainable demand for farmland. When I was young, food was relatively expensive. We ate less food than we typically do now, and we certainly ate less meat. Meat was a rare treat. It is only fairly recently that it has become an everyday food item for many people as the world has become wealthier. The production of meat has also become industrialised, bringing prices down. Like much of our consumption, meat-eating is not evenly spread across the world. Today, the average person in the United States eats over 120kg of meat each year. People in European countries eat between 60kg and 80kg each year. The average Kenyan eats 16kg of meat, and the average person in India, a nation in which vegetarianism is common because of religious beliefs, eats less than 4kg each year.[25]

A piece of meat at our table requires a huge expanse of land for its production. Today, nearly 80 per cent of farmland worldwide is used for meat and dairy production – 4 billion of our 5 billion farmland hectares, an area that would cover both North and South America. Surprisingly, much of this space has no livestock in it at all. It is dedicated to crops like soy, often grown in a different country exclusively as feed for cattle, chickens and pigs. So, the space that livestock actually requires may be unrecognised. Those living in wealthier nations may order meat raised in their country, but some of the feed for those animals will probably have come from tropical nations that are destroying their forests and grasslands to grow feed crops for those animals. It is largely in these tropical nations that the expansion of farmland is still happening, and the world's growing appetite for meat is a leading cause.

Of all the meats, it is beef that is on average by far the most damaging in its production. Beef makes up about a quarter of the meat that we eat, and only 2 per cent of our calories, yet we dedicate 60 per cent of our farmland to raising it. Beef production occupies 15 times more land per kilogram than either pork or chicken. It is simply not going to be possible for every person in the future to expect to eat the amount of beef now consumed by people in the wealthiest nations today. We don't have enough land on Earth to do so.

A wealth of research has already been done to deduce what kind of diet would be fair, healthy and sustainable – both good for people and good for the planet. The universal opinion is that in the future we will have to change to a diet that is largely *plant-based* with much less meat, especially red meat.[26] This will not only reduce the amount of space we need for farmland, and produce fewer greenhouse gases, it will be much healthier for us too. Studies suggest that if we begin to eat a diet with less meat, deaths from heart disease, obesity and some cancers could drop by up to 20 per cent, saving a trillion dollars in healthcare worldwide by 2050.[27]

However, eating meat and raising animals is an important part of the culture, traditions and social life of many people. Meat production also provides livelihoods for hundreds of thousands of people around the world, and in many areas there is no current alternative. How will we make the transition from our current state to a largely plant-based existence? To my mind, this is the second great social change that we will have to undertake over the next few decades. Along with removing fossil fuels from our lives, we will also reduce our dependence on meat and dairy. Indeed, this has already started to happen. Recent surveys show that a third of Britons had either stopped or reduced their meat consumption and 39 per cent of Americans are actively trying to eat more plant-based foods.[28] A similar trend has been found in many other nations. Indeed, I've found in recent years, without taking any sudden decision, I've gradually stopped eating meat. I can't pretend it was overly purposeful nor even that I feel virtuous for having done so, but I have been surprised to realise that I don't miss it. The entire food industry is developing ways to accommodate this trend.

The largest fast food chains and supermarkets are all now experimenting with *alt-proteins*, foods that look, feel and taste like meat or dairy products, but that do not have the animal welfare issues or environmental impacts of livestock farming. Plant-based alternatives to milk, cream, chicken and burgers are now very easy to find, and some of them are remarkable approximations of the original and can offer all the nutrients we need. While soy is a common ingredient in these products, in choosing to eat them ourselves, we are taking the position of herbivore rather than carnivore and so it is far less damaging to the environment than eating animals fed on soy.

At some point, *clean meats* will be arriving on the shelves. These are products grown from genuine animal tissue as independent cell cultures. Since clean meat production does not involve raising livestock, it is very efficient. The cultures are fed on a refined growth medium made from essential nutrients. They don't require much water, energy or space to make, and there are far fewer animal welfare issues.

Further ahead still, there is a possibility of advances in biotechnology that will enable us to use micro-organisms to produce almost any protein or complex organic food to order. Some of these may be produced by adding little more than air and water, and be powered by renewable energy.

At present the cost of producing most of these alt-proteins is still very high since the technology is yet to be refined, and not all are yet proven to be fit for human consumption. Others have been criticised for being overly processed. But some suggest that as soon as they become as cheap to produce as beef, chicken, pork, dairy and, indeed, fish, there will be a revolution in our food supply chains.[29] The bulk of easily substituted foods such as ground beef, sausage meat, chicken breast and milk products may switch to alt-protein production within decades. Even if more specialist items such as prime steaks, fine cheeses and cured delicacies remain produced by traditional methods, the human population would be able to feed itself on far less land, while using far less energy and water, and emitting far fewer greenhouse gases. The alt-protein revolution could prove to be a significant boost to our efforts to become sustainable on Earth.

The UN's Food and Agriculture Organization (FAO) estimates that, with the current rate of improvements in farming efficiency alone, we will reach *peak farm* by about 2040.[30] At that point, for the first time since we invented farming, 10,000 years ago, we may stop taking up more space on Earth. But by radically increasing yields in sustainable ways, regenerating degraded land, farming in new spaces, reducing the meat in our diet, and benefitting from the efficiencies of alt-proteins, we may be able to go much further and start to reverse the land grab. Estimates suggest it could be possible for humankind to feed itself on just half of the land that we currently farm – an area the size of North America. And that would be very valuable, because we have an urgent need for all that freed land. It is the setting for our greatest efforts to increase biodiversity and capture carbon. And the farmers who will have been most affected by the clean, green revolution happening around them, have a pivotal role to play.

Rewilding the Land

At one point, much of old Europe was covered in a deep, dark forest. To the tiny, fledgling farming communities scattered throughout the continent, the forest was regarded as something of an enemy, a barrier to their attempts to establish their meagre fields and feed themselves; a place to fear, haunted by strange spirits and wild beasts. They told fairy tales to their children at night, warning them never to stray into the forest alone. Wolves would have them for supper. The forest would confuse them with its magic, and they would be lost forever. Witches were waiting within. Woodcutters and huntsmen who conquered the forest were cast as heroes. The wild wood, with its relentless growth entombing sleeping princesses and overwhelming empty castles, was the ever-present villain.

The farmers fought the forest with all their might, burning and felling ranks of chestnut, elm, oak and pine, driving it from the river edge and up the valley sides. They killed the wild beasts that lived within it and hung their heads as trophies on the wall. They learnt how to modify the trees, slicing ash, hazel and willow down to the base to create a thicket of long slender trunks, so that they could fashion fencing, thatch and bedposts. Their farms and their numbers grew. Their fear waned. The forest was domesticated.

Deforestation is something we humans do. It is an emblem of our dominance. The relationship between progress and the removal of the forest is so close, there is a recognised model to define it. A nation's *forest transition* describes the deforestation and then *reforestation* that tends to happen in a developing nation over time. When the human population is low and dispersed as small subsistence-farming communities, it is able to do little more than fragment the forest. But this brings wind and light into the woodland, changing its internal environment, and affecting its species composition. The more the forest is fractured, the less able it is to support its original, old-growth community.

As farmers start to trade their produce, a market economy takes over, the farms become businesses and the number and size of fields increases. The value of cultivated land rises quickly, and the remaining forest becomes a target. The great forest is soon reduced to pockets of woodland and stray copses between fields. But, with time, as agricultural practices improve yield, towns and cities attract more and more of the rural population to adopt an urban life, and, increasingly, crops and timber are imported from abroad, there is less need for farmland. Marginal farmland is abandoned first, and the forest begins its return.

Most of Europe had entered the reforestation stage of this transition, in which net forest cover begins to increase, by the Second

World War. The eastern United States, which was stripped of its forests at extraordinary speed upon the arrival of the Europeans, also started to reforest in the first half of the twentieth century. Between 1970 and today, the Western United States, some of Central America, and parts of India, China and Japan have also done so. It has to be noted that a very significant reason all these nations have been able to reforest is because they are, due to globalisation, increasingly importing their food crops and timber from less developed nations. Hence, it is hardly surprising to find that the tropics is still being actively deforested. Many nations in these latitudes, paid for by markets for beef, palm oil and hardwoods in the wealthier parts of the world, are chopping down the deepest, darkest and wildest forests of all – the tropical rainforests. So should they be encouraged to complete their forest transition as fast as possible? Unfortunately, we can't afford to wait. If the forest transition in the tropics runs its course, the loss of carbon to the air and species to the history books would be catastrophic for the whole world. We must halt all deforestation across the world now and, with our investment and trade, support those nations who have not yet chopped down their forests to reap the benefits of these resources without losing them.

That is easier said than done. Preserving wild lands is a very different prospect to preserving wild seas. The high seas are owned by no one. Domestic waters are owned by nations with governments able to make broad decisions on merit. Land, on the other hand, is where we live. It is portioned into billions of different-sized plots, owned, bought and sold by a host of different commercial, state, community and private parties. Its value is decided by markets. The heart of the problem is that, today, there is no way of calculating the value of the wilderness and environmental services, both global and local, that it provides. One hundred hectares of standing rainforest has less value on paper than an oil palm plantation. Tearing down wilderness is therefore seen as worthwhile. The only practical way to change this situation is to change the meaning of value.

The UN's *REDD+* programme is an attempt to do just that.[31] It is a method of giving proper value to the world's last remaining rainforests by pricing the immense amount of carbon they store. That makes it possible to offer the people and governments that keep them in their wild state a payment for doing so, in part funded by carbon offsets. In theory, REDD+ should work. In practice, however, the complications of land ownership and value have raised difficult issues. Indigenous peoples have protested that REDD+ strips the value of the forest down to nothing more than a dollar sign and encourages a new form of colonialism. The money to be made has attracted so-called carbon cowboys from other nations, who swoop in to secure stakes on rainforest land as it gains value. Others fear that in creating a system in which carbon can be offset in the tropics, big industry will use REDD+ as a way to justify their continued use of fossil fuels.

It is a sad fact that when something becomes valuable it will bring out the greed in humankind. As REDD+ learns from its

existing projects in South America, Africa and Asia, the expectation is that it will discover how to improve its approach. We do
need something like REDD+. It is a brave attempt to address a
fundamental undervaluing of nature, and we have to persevere. Its
essential truth is something we all instinctively understand. The
Earth's last forests, rainforests, wetlands, grasslands and woodlands
are, in fact, priceless. They are the carbon stores that we cannot
afford to unlock. They offer environmental services that we cannot
do without. They are home to biodiversity that we must not lose.
How can we come to represent all that in our value systems?

Perhaps we need to change the currency. The danger with
pricing nature purely on the amount of carbon it captures and
stores is that carbon then becomes the only thing that matters to
us. It oversimplifies nature's value to us, but, worse, it may lead
us to imagine that fast-growing eucalyptus plantations are as
valuable as biodiverse forest. We may choose to use the farmland
no longer required for food production purely for monocultures of
bioenergy crops rather than restoring woodlands. Carbon capture
and storage is extremely important, but it isn't everything. It won't
stop the sixth mass extinction. To create a stable and healthy world,
it is biodiversity that we ought to be cherishing. After all, if we
increase biodiversity, we will, by default, maximise carbon capture
and storage, since the more biodiverse a habitat, the better it does
that job. What would a world look like in which biodiversity was
properly valued and landowners were encouraged to increase it,
wherever and however they can?

It would be magical. Primary rainforest, old-growth temperate forest, intact wetlands and natural grasslands would suddenly become the most valuable real estate on Earth! The owners of these wild lands would be rewarded for continuing to protect them. Deforestation would immediately halt. We would quickly realise that the best place to plant palm oil or soy is not on land occupied by virgin rainforest, but on land that was deforested years ago – after all there's plenty of it.

We would be encouraged to find ways to use pure wilderness without reducing its biodiversity or its ability to capture carbon. And such practices do exist. The respectful surveying of virgin rainforest for unknown organic molecules that might lead to new cures for disease or industrial materials or foodstuffs, could be acceptable – provided that local communities give their consent and the subsequent commercial gain of these items brought back income to those who safeguarded the forest. Sustainable logging,[32] in which select trees are felled and carefully removed at rates that mimic the natural turnover of a forest, would be permitted, for this has been shown to preserve biodiversity.[33] Ecotourism, which enables all of us to experience the wonders that are being protected, can bring a great deal of income to wild places without significant impact. Indeed, the more wilderness there is in the future, the more dispersed the tourists could be.

And there would be a great drive to expand and regenerate all lands that adjoin pristine wilderness. The best people to lead these initiatives would be the local and indigenous communities that live in and around our wildest lands. Experience in conservation projects has shown that positive change will only last for the long term if local communities are fully involved in developing the plans and directly feel the benefit of rising biodiversity. One story from Kenya demonstrates this. The Maasai are herders, who, for hundreds of years, have grazed their cattle and goats on the Serengeti plains alongside the wildlife. They don't eat the wild animals about them. They even tolerate the local predators taking a few of their cattle each year. As Kenya has developed, the population of the Maasai grew. Overgrazing by domestic herds subsequently started to become a problem. Their neighbours, the wild animals, began to disappear. In response, Maasai families teamed up to create *conservancies* with the aim of bringing the wildlife back. They agreed to herd their cattle in such a way that they promote a mosaic of vegetation, attracting a greater number and variety of herbivores, and hence predators. As the conservancy rewilds, the families grant licences for low-impact safari lodges to operate on their land. The model then starts to work. The more the wildlife returns, the more people want to visit the safari lodges and the more the Maasai community gains. After only a few years in operation, some Maasai families have started to actually reduce their cattle herds in order to boost wildlife further. When I visited these conservancies in 2019, those of the younger generation of Maasai were quick to explain that they are coming to value wild herds over their domestic herds. Now Maasai communities in adjacent lands, seeing the successes of their neighbours,

are adopting the conservancy model too. Within a few decades it may be possible, through a network of protected areas connected by wildlife corridors, to have wild grasslands that stretch from the shores of Lake Victoria to the Indian Ocean, purely because biodiversity has been found to be of genuine practical value.

There is hope that the wild can return even to land that was first cultivated in Europe long ago. As the demand for space for food production wanes, European governments are indicating that they may alter the subsidies they pay out to farmers to reward them for using land in a way that maximises biodiversity and carbon capture.[34] This new regime could trigger a remarkable response on millions of hectares of European farmland. We might expect to see hedgerows come back to replace the fencing. There could be an explosion in agroforestry, in which crops are grown beneath trees. Ponds and waterways could be restored on farms. Pesticides and fertilisers, both of which damage biodiversity, would start to lose their attraction. Farmers may instead plant crops that draw animal pests away from the food, and adopt regenerative techniques to make their soils naturally rich.

This wild approach to farming may find its strongest advocates among the meat producers. As people adopt a *plant-based diet*, perhaps they will become more selective of the little meat they buy, going for quality rather than quantity. People may seek out beef, lamb, pork and chicken raised in ways that capture carbon and promote wildlife. In response, animal farmers may choose to switch from the intensive feed-lots and battery farms using imported feed to practices such as *silvopasture*, in which animals are raised year-round within growing woodlands. The volume of production is much lower than intensive farming, but the planet-friendly product could carry a premium. The trees in the fields more than compensate for the emissions of the animals and provide the shade and shelter needed to improve their health and yield. The animals in return fertilise the soils and keep unwanted plants at bay.

Silvopasture works so well simply because it replicates a natural state. In prehistoric times, long before Europe was densely forested, it was a land of wood pasture, a mosaic of wildwood frequently broken up by meadows. This landscape was created by the browsing of a wild community of giant, fierce wild cattle known as aurochs, wild horses called tarpan, herds of European bison, elk and wild boar – all the animals featured on the walls of caves in France. It is a natural community that two adventurous livestock farmers have been trying to mimic in the south of England.

In 2000, Charlie Burrell and Isabella Tree took a leap of faith on their 1,400-hectare farm, the Knepp Estate.[35] Faced with bankruptcy due to the increasing costs of machinery and agrochemicals on their marginal land, they decided to abandon the commercial farming they'd been practising all their lives and return their farm to the wild. They broke open the fields, selected breeds of cattle, pony, pig and deer that best replicated the mix of species that would have been present on the land thousands of years before, and let them mingle and roam freely, year-round, without supplements. In

mixing herbivores naturally like this, they began to mimic interactions in the wild. There, zebra and wildebeest work together
to graze the grasslands. The zebra take the tougher, taller grass,
leaving the wildebeest with the softer, leafy grass they are able
to digest. Studies have shown that when cattle are grazed with
donkeys in this way, they can gain significantly more weight as a
result of feeding together than when kept apart. This and many
other complementary effects operate in a wild habitat. They are
instrumental in determining the future direction of a landscape,
and they began to transform the farm at Knepp. The animals,
acting together like the wild stock of prehistoric England, started
to turn the uniform fields into new marshes, thickets, heaths and
woodlands. As a result, the biodiversity of the farm exploded.
Within only 15 years, it became one of the best places in England
to find a host of rare, native plants, insects, bats and birds.

Charlie and Isabella's *wildland farm* still produces food. Each
year, they judge the number of animals that the changing landscape
can support and harvest the surplus. They are, in effect, doing the
job of a top predator.

Knepp is not a conservation project, in that it doesn't have a goal
or target species it wishes to benefit. It simply lets the animals be
the drivers of the landscape, and they are doing an excellent job. In
addition to its record-breaking diversity, the farm is now sequestering
tonnes of carbon in its enriching soils and its changing waterways
are mitigating flooding downstream. Arguably, the Knepp Estate, a
working livestock farm, is now the closest approximation of ancient,

wild Britain to be found anywhere. There are plenty of people eager to visit. Eco-safaris and wild camping have added to the income from meat and subsidies, and the farm is finally profitable.

Wildland farms could become commonplace in an era in which biodiversity is appropriately rewarded. Any mix of animals that would serve as a proxy for the native community would lead to the habitat reverting towards its natural state. If tourism is not an option to supplement income, perhaps farmers could turn to other complementary livelihoods such as clean energy generation. The giant wind turbines being manufactured today could stand over an open grassland or even, as now demonstrated in Germany, above a forest, without disturbing the developing wilderness. The animal farmers of the future could, with the right support, be more than food producers. They could become soil engineers, carbon traders, foresters, tour guides, energy suppliers and curators of the wild – custodians skilled at harvesting the natural potential and sustainable value of their land.

Conceivably, with the right motivation, the wildland farm approach could scale up to change whole landscapes. With biodiversity, it is almost always the case that a greater area brings even greater rewards. If neighbouring landowners agree to share their revenue, they could unite, creating huge borderless parks, similar in many ways to the Maasai conservancies. Communities of landowners are already uniting hundreds of thousands of hectares in projects to increase biodiversity on the Great Plains of North America and the steep, forested valleys of the Carpathians in Europe.[36] It is possible.

When working on a large scale, the opportunity arises for the most spectacular and controversial of rewilding ambitions – the reintroduction of large predators. In a world in which biodiversity gain and carbon capture are rewarded, it may make sense to do this, given enough space, due to the benefits of something called the *trophic cascade*. The most famous example was recorded in Yellowstone National Park upon the reintroduction of wolves in 1995. Until the wolves came back, the large deer herds spent long hours browsing the shrubs and saplings that were growing in river valleys and gorges. When the wolves arrived, that stopped, not because the wolves ate lots of deer, but because they scared all the deer. The routine of the deer herds changed. Now they moved frequently and did not remain in the open for long. Within six years, the trees grew back, shading the water, allowing fish to gather out of sight. Aspen, willow and cottonwood thickets sprouted on the floors and sides of the open valleys. The numbers of woodland birds, beaver and bison increased. The wolves hunted coyote too, so populations of rabbits and mice did better and so fox, weasel and hawk numbers increased. Finally, even the bears grew in number, as they benefitted from scavenging the carcasses of wolf kills. In the autumn, they feasted on the berries of trees and shrubs that would otherwise never have come into fruit.[37]

The conclusion is clear: to gain biodiversity and capture carbon in a landscape such as Yellowstone, just add wolves. This thinking is active in the minds of Europeans now planning to deal with the 20–30 million hectares of abandoned farmland expected to be created by the continent's continuing forest transition by 2030. This is an area the size of Italy. If forests are about to return to the farms by natural regrowth, it would be better for them to be as biodiverse and carbon efficient as possible. The return of the wild is becoming a practical policy option for governments which understand the true value of nature and its contribution to stability and well-being.

All the incentives are set to bring about a much wilder world by the end of this century than there was at the beginning. Sceptics need only look at the nation of Costa Rica to understand what is possible with the correct motivations. A century ago, more than three-quarters of Costa Rica was covered with forest, much of it tropical rainforest. By the 1980s, uncontrolled logging and the demand for farmland had reduced the country's forest cover to just one-quarter. Concerned that continual deforestation would reduce the environmental services of its wild lands, the government decided to act, offering grants to landowners to replant native trees. In just 25 years, the forest has returned to cover half of Costa Rica once again. Its wild lands now provide a significant component of the nation's income and have a central role in its identity.

Just imagine if we achieved this on a global scale. A study from 2019 has suggested that the return of the trees could theoretically absorb as much as two-thirds of the carbon emissions that remain in the atmosphere from our activities.[38] The rewilding of the land is within our gift, and it is undoubtedly a valuable thing to do. Creating wild lands across the Earth would bring back biodiversity, and the biodiversity would do what it does best: stabilise the planet.

Planning for Peak Human

Up to this point, this vision has been concerned with reducing the footprint of our consumption and enabling the wild to return in as many meaningful ways as possible. If we wholeheartedly embrace all these measures, we will certainly have far less overall impact on the Earth. Even those most fortunate in life, who presently have the biggest footprints, will be closer to a sustainable existence. So the impact of our entire species would be more equally distributed. However, to secure the grand ambition of the Doughnut Model, a stable world in which everyone gets a fair share of its finite resources, we have to take into account the level of our own population.

When I was born, there were fewer than 2 billion people on the planet – today there are almost four times that number. The world's population is continuing to grow, albeit at a slower pace than at any time since 1950. At current UN projections there will be between 9.4 and 12.7 billion people on Earth by 2100.[39]

In the wild, animal and plant populations in any one habitat remain roughly stable in size over time, in balance with the rest of the community. If too many are alive at once, each individual will find it harder to get what it needs from the habitat, and a few will die or leave. the habitat entirely. If too few are born, there will be more than enough to go round. So they will breed well and the species will reach its full potential once more. Increasing slightly, decreasing slightly, the population of each species oscillates about a number that the habitat can sustain indefinitely. This number – the *carrying capacity* of an environment for a particular species – represents the very essence of balance in nature.

What is the human carrying capacity of the Earth? Despite reasoned proposals and fearful warnings from great thinkers throughout history, we have never yet met our natural ceiling. We always seem to invent or discover new ways of using the environment to provide more of the essentials – food, shelter, water – for ever more people. Indeed, it is more impressive than that. We effortlessly support far more than the essentials – schools, shops, entertainment, public institutions – even as we increase our population at an extraordinary speed. Is there nothing to stop us?

The catastrophe unfolding around us surely suggests that there is. The loss of biodiversity, the changing climate, the pressure on the planetary boundaries, everything points to the conclusion that we are finally fast approaching the Earth's carrying capacity for humanity. Each year, since 1987, an Earth Overshoot Day has been announced – an illustrative date in the calendar on which humankind's consumption for the year exceeds the Earth's capacity to regenerate those resources in that year. In 1987 we were overshooting the Earth's resources by 23 October. In 2019, we were doing so by 29 July. Humankind now uses up the equivalent of 1.7 times what the Earth can regenerate in a single year.[40] Whilst 60 per cent of this figure is the result of our carbon emissions footprint, it gives a clear indication of how excessive our demand on nature has become. This overshoot is the nub of our unsustainability – we are distorting the Earth's carrying capacity by eating into the capital of its resources. The catastrophe ahead is what happens when the Earth calls in our overdraft.

By reducing the impact of our consumption in all the ways outlined above, we will effectively raise the Earth's carrying capacity once again, so that more of us can share this planet. But, clearly, to give everyone the fair share they deserve and improve the lives of all as the Doughnut Model demands, it is important that human population growth does level off. Happily, the evidence shows that improving the lives of everyone does exactly that.

Demographic transition is a term used by geographers to describe the path that nations move along during their economic development. It has four stages, though many nations are currently yet to complete all four. Progress through the transition is marked by changes in birth and death rates. As countries move along the path, they experience a boom in population, followed by a levelling off to a stable plateau – a maturation, as it were. Japan made its way through this transition during the twentieth century. For millennia, Japan had been in Stage 1 of the transition – a preindustrial society, based upon agriculture, prone to the disasters

Model of Demographic Transition

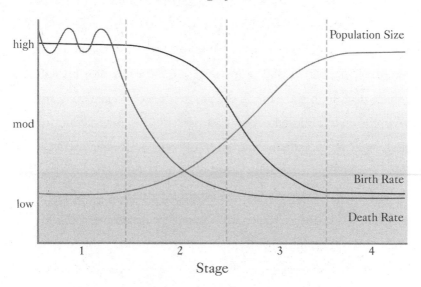

high

mod

low

Population Size

Birth Rate

Death Rate

1　　　　　2　　　　　3　　　　　4

Stage

of drought, floods and infectious diseases. The birth rate was high but the death rate was also high, so the population changed very little, growing slowly over the centuries. By 1900, however, Japan was urgently industrialising. The Japanese governments of the nineteenth century, determined to prevent being colonised by the European nations, had embarked on a policy of 'wealthy country / strong military'. Huge amounts of investment in science, engineering, transport, education and agriculture transformed Japanese society. Industrialisation took Japan into Stage 2, in which birth rate stays high, but death rate drops. The improved food production, education, healthcare and sanitation of industrialisation brought a sharp decline in the nation's death rate. Since women were still having as many children as they had always had – usually four, five or six – Japan's population started to balloon. It doubled between 1900 and 1955, to 89 million people.

Immediately after the Second World War, as a defeated power overseen by the Allies, Japan was forced to abandon its military ambitions and rebuild by aligning with the global economy. As the Great Acceleration began, creating a surge in the demand for consumer products such as washing machines, televisions and cars, Japan was well-placed to become a dedicated supplier of technology. A so-called miracle growth occurred between the early 1950s and early 1970s, in which cities grew fast, incomes rose, education improved and aspirations grew. But, critically, during this period the birth rate suddenly dropped. By 1975, the average family had only two children. Many aspects of life were better for most, but they were also more expensive. There was less room, money, time to raise families – and there was less incentive for big families because child mortality had dropped with improvements in diet and healthcare. Japan was moving through Stage 3 in which the death rate remains low, but the birth rate falls. The population boom started to dwindle as the family size decreased. The growth curve was coming to a peak.

By 2000, the population of Japan was 126 million. That is what it is today. The population has levelled out. Japan is in Stage 4 of the transition – both birth rate and death rate are low, meaning that, once again, they cancel each other out and the population remains stable. The population explosion in Japan was a temporary, one-time event, ultimately checked by the advances in society brought by the Great Acceleration.

This four-stage demographic transition is happening today to all nations across the globe. The huge leap in human population during the twentieth century was the result of hundreds of nations travelling through Stages 2 and 3 of the demographic transition. It is possible to chart such a transition in the population of the whole world. The rate at which the world population has grown each year peaked as early as 1962, and since then, it has broadly been dropping year by year. This implies that the transition of the world population on average from Stage 2 to Stage 3 happened at that point. Since that time the average family size on Earth has halved. In the early 1960s, women would typically have five children. Today, the average is 2.5. The world is approaching the end of Stage 3.[41]

Of course, the big question is: when does the world settle into Stage 4? At what moment will the world population do what Japan's did, and peak? That will be a historic occasion – the day that those who study population, the demographers, call *peak human*, the moment our population stops growing for the first time since farming began 10,000 years ago. It will be a milestone on our journey to regain our balance on Earth.

However, the reality is that, even upon reaching Stage 4 globally, it will take a long time for our population to get to its summit, due to what Swedish social scientist Hans Rosling called the 'inevitable fill-up'.[42] Firstly, family size must drop sufficiently for us to reach *peak child* – the point from which the number of children on Earth stops increasing. Then we have to wait for this largest-ever generation of children to pass through their twenties and thirties – the time in which they will have children – before the population starts to plateau. In essence, it is only when we get past 'peak mother' with as low a family size as possible, that the population will stop growing.

Added to this, the total number of people on Earth is further inflated by what is, on the face of it, a positive trend of which I am certainly a part – a rising life expectancy. As nations progress through the demographic transition, life expectancy climbs rapidly. In Stage 1, when child mortality, disease and poor diets are a normal part of life, people live around 40 years. By Stage 4, they live twice as long. In fact it is predicted that by mid-century there will be more than twice as many persons over 65 years old as there are children under five. The inevitable fill-up gives our population huge momentum – the opposite of the inertia it experienced at the start of the boom a century ago – and this momentum makes it unlikely that we will get to peak human within this century. In 2019, the United Nations' Population Division published its latest projections for global human population. They indicated that if the globe's demographic transition plays out as we expect it to, the human population will peak early in the twenty-second century at about 11 billion people, 3.2 billion more than today. Due to the nature of the curve, there is relatively little increase in the population from 2075, a moment only 55 years into the future. But is there a way in which we can encourage it to peak still sooner and lower?

China thought it had the answer in 1980 when it put in place its one-child policy. Quite aside from the moral issues, the difficulty of administering the policy, and the social and cultural disruption associated with it, there is little evidence such an approach works any faster than economic development. In the time that China's average family size dropped from six children to just over one, neighbouring Taiwan experienced a greater drop without following a one-child policy, purely as a consequence of going through its natural transition at speed.[43] It seems that the best way to stabilise the population is to support nations that seek to speed up their demographic transition. In practical terms, this means helping the least developed nations to achieve the ambitions in the Doughnut Model as fast as possible – supporting people as they raise themselves out of poverty, building healthcare networks, education systems, better transport and energy security, making these nations attractive to investment – anything, in fact, that improves the lives of people. Among all these social improvements, one in particular is found to significantly reduce family size – the empowerment of women.[44] Wherever women have the vote, wherever girls stay in school for longer, wherever women are in charge of their own lives and not dictated to by men, wherever they have access to good healthcare and contraception, wherever they are free to take any job and their aspirations for life are raised, the birth rate falls. The reason for this is straightforward – empowerment brings freedom of choice and when life offers more options for women, their choice is often to have fewer children. The faster and more fully women are empowered, the quicker a nation will move through Stage 3 and on to Stage 4.

This empowerment can take many forms. In parts of rural India, only 40 per cent of girls attend school beyond the age of 14. The distance to a high school is often so great, that teenage girls find they cannot get back and forth from school in the day and still have time to do the household tasks asked of them. Several state governments and charity projects have provided hundreds of thousands of free bicycles in response, and the freedom these provide has radically improved the attendance of girls. It is now common to see girls cycling in groups between the fields of rural India, able to finish their education.

Research at Austria's Wittgenstein Centre has demonstrated how dramatically a strong multinational effort to raise the standards of education across the world would change the course of human population growth.[45] In one of their forecasts, they calculated what would happen if education systems in the world's poorest nations improved as quickly in this century as they did in the fastest developing nations of the last century. At this fast-track level, peak human occurs as soon as 2060, at a level of 8.9 billion people. This is an astonishing revelation – by simply investing in social and education systems, we may be able to reduce the peak of the human population by more than 2 billion people, and bring it about 50 years earlier. Even if there are some errors in the assumptions, this model combined with real-world examples surely gives us a clear path to assist the prospects of all humankind by eagerly improving the lives of those with greatest need.

Raising people out of poverty and empowering women is the fastest way to bring this period of rapid population growth to an end. And why wouldn't we want to do these things? This is not just about the numbers of people on the planet. This is about committing to a fair and just future for all. Giving people a greater opportunity in life is surely what we all would want to do anyway. It's a wonderful win–win solution, and this is a repeating theme on the path to sustainability. The things we have to do to rewild the world tend to be things that we ought to be doing regardless.

* * *

When we finally achieve peak human, it will be a significant event. Yet it is not necessarily the end of the journey. There is

some evidence that the demographic transition has a Stage 5. Japan's population is now in decline. The forecast is that by the 2060s it will reach 100 million people, about the same number as there was in Japan in the 1960s. As it declines, Japan's population will also age – there will be an increasing proportion of older people. Economically, this poses a significant problem. A reducing working population will be asked to support an increasing number of elderly people. Indeed, this process has already started, and, as one of the first countries in the world to tackle this fifth stage of transition, there is a great deal of soul searching in Japan about what to do. The current imperative for endless growth in GDP tempts politicians to call for more babies to provide more future workers, or demand that retired Japanese go back to work to help the tax burden of those in their middle age. Others suggest that Japan, if anywhere, should be able to bring in robot and Artificial Intelligence to help maintain the economy. Should we move to a world economy that is less reliant on growth, one would hope that the relentless push for economic performance may ease, and Japan, followed by all other nations, will find a comfortable equilibrium with fewer people in a more mature and dependable world.

By working hard now to improve as many people's lives as much as possible, the most optimistic models suggest that human population may return to the level it is today by the end of this century. After that, perhaps our population will continue to reduce at a gentle rate, the global society demanding less from our world and helping to meet its needs with technological solutions, much as it always has done.

However, we have a very long and formidable journey to make to get to that point without catastrophe. The inevitable fill-up, the increase in human numbers still to come over many years, brings another inevitability – the decisions we make today are even more critical. We need all to align and work hard to give everyone a fair and decent standard of living as soon as possible.

Achieving More Balanced Lives

A revolution in sustainability, a drive to rewild the world and initiatives to stabilise our population would realign us as a species in harmony with the natural world about us. How would it affect our own, individual lives? In a thriving, sustainable future, we would follow a largely plant-based diet, filled with healthier alternatives to meat. We would use clean energy for all our needs. Our banks and pension funds would only invest in sustainable business. Those of us that choose to have children would be likely to have smaller families. We would be able to choose wood products, foodstuffs, fish and meat thoughtfully, informed by the detailed information available with every purchase. Our waste would be minimal. The little carbon our activities still emit would be offset automatically within the purchase price, funding rewilding projects all over the world.

In truth, it would be easier for us, in this potential future, to live a life in balance with the natural world than it is today. Business and political leaders will have been compelled to build products and societies that help all of us to have a lower impact. Take for example, the treatment of waste. I can remember a time before the disposable society we have today, when we repaired and reused, when we had little or no plastics, and food was a precious commodity. The present habit of throwing everything away, even though, on a finite planet there is of course no such thing as 'away', is a relatively new thing. Aside from the fact that waste is a waste, when it accumulates it often becomes damaging. The living world faces the same problem, and we will, once again, be wise to copy its solutions. In nature, the waste from one process becomes the food for the next. All materials are reused in cycles, involving many different species, and almost everything is ultimately biodegradable.

Those studying possibilities for a *circular economy*, such as the researchers at the Ellen MacArthur Foundation,[46] are looking for ways to bring the same logic and efficiencies into our societies. The key to the circular mindset is to imagine replacing the current take-make-use-discard model of production with one in which raw materials are thought of as nutrients that must be recycled, just as nutrients are in nature. It then becomes clear that we humans are essentially engaged in two different cycles. Anything that is naturally biodegradable – food, wood, clothes made from natural fibres – is part of a biological cycle. Anything that is not – plastics, synthetics, metals – is involved in a technical cycle. The raw materials in both cycles – the carbon fibres or titanium, for example – are elements that need to be reused. The cleverness comes in designing ways to do so.

In the biological cycle, food waste is the key component. As we have seen, food production at present entails deforestation, the use of fertilisers and pesticides and the use of fossil fuels in its transportation. Food is also expensive and many people around the world still struggle to afford a healthy diet. And yet, globally, we lose and waste one-third of all the food we produce.[47] In poorer countries, with less infrastructure, the bulk of the waste occurs before it gets to the shops due to harvest losses, damage and poor storage. In wealthier countries, it mainly occurs after harvest. Some is cast aside due to perceived imperfections, some discarded as surplus because of over-ordering, and a large amount simply not eaten and thrown away. In a more sensible world, infrastructure and storage would be improved. Businesses might feed the waste to livestock, or send it to insect farms that breed flies for fish and animal food. They might use the more fibrous waste, such as nut shells, as bioenergy fuel in combination with wood scraps from the timber industry to create heat and electricity. In doing such things, they could capture the escaping carbon and store it. They may even bake the waste in an oxygen-free environment to create *biochar*, a charcoal-like mass that can be used as a building material, a low carbon fuel or an additive to soils that enriches them and locks carbon back away beneath the surface.

In the technical cycle, many of the circular efficiencies come from coordinating the design of products. The companies making items from plastics, synthetics and metals could build them to last, rather than to work for only a few years. They could build components in such a way that they are easily removed, disassembled, re-formed and updated. Manufacturing would have to become far more standardised, so that components could be made by multiple providers and swapped. All product lines would have to have a plan for the clever sourcing and the onward destinations of all the elements involved. Some believe that the cyclical approach would drive new relationships between customers and companies, such that customers merely rent washing machines and televisions from a manufacturer, as they do phone handsets today, though with a far greater emphasis on mending and recycling.

In both cycles, any materials or chemicals that cannot be recycled, or that are inherently dangerous to the environment, would be removed from the economy over time. Chief among these are the manmade hydrofluorocarbons (HFCs) currently residing in the world's refrigerators and air conditioners. If these were released from machines at the end of their life, they would add the equivalent of 100 gigatonnes of carbon dioxide to the atmosphere in greenhouse gases. An international agreement in 2016 has already paved the way for their safe transformation into chemicals that do not cause global warming.[48]

The circular ambition is to create a world that has no pollution – no plastics floating in the sea, no toxic gases emitted from industrial chimneys, no burning rubber tyres, no oil slicks. It would be a world that might even undo the wastefulness of today. Our landfill sites could become open-cast mines for companies paid handsomely to dig out nutrients for the circular economy. The microplastics circling in gyres in the ocean could be retrieved and combined to build ocean farmsteads. By changing our approach to the use of our resources, a growing number of people believe that humanity could eradicate waste and come to mimic nature's cyclical approach.

What of the places in which we carry out our lives? By 2050, 68 per cent of the world population is predicted to live in cities. At one time, cities were regarded by environmentalists as the scourge of the planet, clogged with energy-guzzling traffic and pollution, their inhabitants' endless need for products and materials creating a dirty footprint across the world. But they have come to appreciate that, due to the high densities of people in cities, the urban environment holds great potential for sustainability. City planners are learning to make their cities friendly to pedestrians and cyclists. They can build in efficient, low-carbon public transport. Some cities, like Copenhagen, are installing systems of centralised district heating which draws its heat energy from geothermal plants or the waste produced in the city itself. The big, expensive buildings at the heart of a city can be required to meet high standards of insulation and energy efficiency. All of this means that a city-dweller's carbon emissions are now often significantly lower than those of someone living in the countryside.

There is a huge incentive for the world's great cities to go much further. In a global market, city mayors understand that they are in competition with cities all over the world for the best talent. One of the most effective ways to attract people to a city is to make it as green and pleasant as possible. As well as providing spaces for leisure, urban plant life has been shown to cool cities, purify the air and improve the mental well-being of the city-dwellers. As a result, cities are welcoming nature by extending parkland, building avenues and encouraging green roofs and walls covered by cascades of plants. Paris is currently adding 100 hectares of green space to its buildings' rooftops and walls. In several Chinese cities, wetlands are being created at the margins of city rivers to soak up seasonal flooding and provide citizens with more natural space. London has declared itself to be the world's first National Park City with a plan to turn over half of its area into natural spaces and make the lives of Londoners greener, healthier and wilder.

The city state of Singapore is intent on transforming itself into a city within a garden. All new buildings are asked to replace the greenery lost on the ground due to their construction with an equivalent amount of plant life above ground. As a result, the city has dozens of buildings designed specifically to be covered in plants, including a hospital that is reporting better recovery rates among its patients as a result of the greenery. Singapore is linking all its parklands with green corridors and has turned 100 hectares of prime land on its shoreline into a water reservoir and garden featuring a grove of 50m artificial supertrees that power themselves with solar panels, irrigate the gardens with the water they have collected and filter the air.

Biologist Janine Benyus, co-founder of the Biomimicry Institute, wishing to provoke the new green approach to city planning, has

set all cities a challenge. She suggests that, since a city occupies space that was once natural habitat, it should at least equal that habitat in terms of the environmental services it once provided – the solar energy it generated, the fertility it added to its soils, the volume of air it cleaned, the water it produced, the carbon it captured and the biodiversity it hosted. Architects appear keen to take on her challenge. The best sustainable buildings being built today are net generators of renewable energy, they purify the air around them, treat their own wastewater, create soil from sewage and offer permanent homes for an abundance of animals and plants. In the future it may be possible for cities to give back rather than just take.

* * *

Give and take, that is the essence of what balance is all about. When humankind as a whole is in a position to give back to nature at least as much as we take, and repay some of our debt, we will all be able to lead more balanced lives. There are examples across the world of this new thinking right now. If every nation were to set profit, people and planet targets for itself as New Zealand does, offer a standard of living for its population as high as Japan's, embrace the renewable revolution like Morocco, manage its seas like Palau, farm plants as efficiently and sustainably as some are doing in the Netherlands, eat meat rarely like the people of India, encourage the wild to return as Costa Rica has, and build nature into its cities like Singapore, the whole of humanity would be able to achieve a balance with nature. But it will take every nation, and those with the greatest footprints to make the biggest changes. It won't work if some countries make the transition and others don't.

There is some resistance at present. It is all too easy when contemplating sustainability to focus on what we lose and miss what we gain. But the reality is that a sustainable world is full of gains. In losing our dependence on coal and oil and by generating renewable energy we gain clean air and water, cheap electricity for all, and quieter, safer cities. In losing rights to fish in certain waters, we gain a healthy ocean that will help us combat climate change and ultimately offer us more wild seafood. In removing much of the meat from our diet, we gain fitness and health and less expensive food. In losing land to the wild, we gain opportunities for a life-affirming reconnection with the natural world both in distant lands and seas and in our own local environment. In losing our dominance over nature, we gain an enduring stability within it for all the generations that will follow.

Everything is set for us to win this future. We have a plan. We know what to do. There is a path to sustainability. It is a path that could lead to a better future for all life on Earth. We must let our politicians and business leaders know that we understand this, that this vision for the future is not just something we *need*, it is something, above all, that we *want*.

Our Greatest Opportunity

I was born in another time. I don't mean this metaphorically, but literally. I arrived in this world during a period geologists call the Holocene, and I will leave it – as will every one of us alive today – in the *Anthropocene*, the time of humans.

The term Anthropocene was proposed in 2016 by a group of eminent geologists. Dividing the Earth's history into named periods has long been geological practice. Each is recognised by characteristics that distinguish the rocks of that particular age from all others – the absence of some fossil species that had flourished earlier and the appearance of new ones.

That will certainly be the case in the rocks that are forming today. Not only will they contain fewer species than the rocks that preceded them but they will contain markers that are completely new – fragments of plastic, plutonium from nuclear activity, and a worldwide distribution of the bones of domesticated chickens. The geologists suggested that this new epoch might begin in the 1950s and that it should be called the Anthropocene, since it is the human species more than any other that is determining its character.

What for the geologists was a name produced by scientific routine has now, however, become to many others a vivid expression of the alarming change that now faces us. We have become a global force with such power that we are affecting the entire planet. The Anthropocene, in fact, could prove to be a uniquely brief period in geological history and one that ends in the ultimate disappearance of human civilisation.

It need not be so. The advent of the Anthropocene could yet mark the beginning of a new and sustainable relationship between ourselves and with the planet. It could be a time in which we learn how to work with nature rather than against it, a time in which there would no longer be any great distinction between the natural and the managed, for we would become the attentive stewards of the entire Earth, calling upon nature's extraordinary resilience to help us bring its biodiversity back from the brink.

In the end, the question of which version of the Anthropocene is about to unfold is up to us. Human beings may be ingenious but they are also quarrelsome. Our history books have been dominated by stories of wars, of struggles for dominance between nations. But we cannot continue in this way. The dangers that now face the Earth are global and can only be dealt with if nations sink their differences and unite to act globally.

There are in fact precedents of our managing to do so. In 1986, the whaling nations of the world got together and decided that the slaughter of whales of all kinds had to end if these extraordinary and wonderful animals were not to be exterminated.

Some delegates may have agreed to stop the hunt because whales were, by then, so reduced in numbers that it was no longer economic to pursue them. But others certainly did so because of pleas from conservationists and scientists. The decision was by no means unanimous. And there are still continuing arguments. But in 1994, 50 million square kilometres of the Southern Ocean was declared an International Whale Sanctuary. Today, as a result of these restrictions, whales have increased to numbers that have not been seen in living memory. And an important and influential factor in the complex workings of the ocean restored to something like its proper position.

In central Africa, where in the 1970s only 300 mountain gorillas survived, cross-border agreements were eventually made between a number of African nations and now there are over a thousand of these magnificent creatures, thanks to the hard work and bravery of generations of local rangers.

So it is within our power to come together internationally, if we want to. Now, however, we must make agreements that apply not just to a single group of animals but to the whole of the natural world. It will take the labours of countless committees and conferences, and the signing of innumerable international treaties. The work has already started, organised by the United Nations. Huge conferences involving tens of thousands of people are at work. One series is dealing with problems concerning the alarming rate that our planet is warming which could have such widespread and devastating consequences. Another series is charged with protecting the biodiversity on which the whole interconnecting web of life depends.

The task could hardly be more daunting and we have to support it in every way we can. We have to urge our politicians, locally, nationally and internationally, to come to some agreement and sometimes subordinate our national interest in support of the bigger and wider benefit. The future of humanity depends upon the success of these meetings.

We often talk of saving the planet, but the truth is that we must do these things to save ourselves. With or without us, the wild will return. Evidence of this is no more dramatic than that to be seen in the ruins of Pripyat, the model city that had to be abandoned when the Chernobyl nuclear reactor exploded. When you step outside the dark and empty corridors of one of its deserted apartment blocks, you are greeted by a most surprising sight. In the 34 years since the evacuation, a forest has taken over the deserted city. Shrubs have broken up the concrete and ivy pulled apart the bricks. Roofs sag under the weight of accumulating vegetation, and saplings of poplar and aspen have burst through the pavements. The gardens, the parks and the avenues are now shaded by the canopies of oaks, pines and maples, 20 feet above the ground. Beneath, there is a strange under-storey of unkempt ornamental roses and fruit trees. The football field, which 34 years ago served as a landing pad for military helicopters sent to evacuate the city's inhabitants, is now covered by a thicket of young trees. The wild has reclaimed its territory.

The land including the town and the ruined reactor has now been designated a sanctuary for animals that are rare elsewhere. Biologists have placed camera-traps at the windows of the town and recorded images of thriving populations of foxes, elk, deer, wild boar, bison, brown bear and racoon dogs. Some years ago, a few individuals of the nearly extinct Przewalski's horse were released there, and their numbers are now increasing. Even wolves have colonised the area, safe from the guns of hunters. It seems that, however grave our mistakes, nature will be able to overcome them, given the chance. The living world has survived mass extinctions several times before. But we humans cannot assume that we will do the same. We have come as far as we have because we are the cleverest creatures to have ever lived on Earth. But if we are to continue to exist, we will require more than intelligence. We will require wisdom.

Homo sapiens, the wise human being, must now learn from its mistakes and live up to its name. We who are alive today have the formidable task of making sure that our species does so. We must not give up hope. We have all the tools we need, the thoughts and ideas of billions of remarkable minds and the immeasurable energies of nature to help us in our work. And we have one more thing – an ability, perhaps unique among the living creatures on the planet – to imagine a future and work towards achieving it.

We can yet make amends, manage our impact, change the direction of our development and once again become a species in harmony with nature. All we require is the will. The next few decades represent a final opportunity to build a stable home for ourselves and restore the rich, healthy and wonderful world that we inherited from our distant ancestors. Our future on the planet, the only place as far as we know where life of any kind exists, is at stake.

Glossary

Alt-proteins (alternative proteins) A general term that covers plant-based and food-technology alternatives to regular animal protein, for example proteins derived from grains, legumes, nuts, seeds, algae, insects, micro-organisms or indeed *clean meat*. Since these proteins do not involve large-scale livestock or fish production, the expectation is that their production will have a much smaller environmental footprint. In addition, they will have fewer animal welfare issues.

Aquaculture (fish farming) – The breeding, rearing and harvesting of fish, shellfish, algae and other organisms in water environments. There are two main categories, marine and freshwater.

Anthropocene – A proposed geological age or, more technically, epoch, viewed as the period during which human activity has been the dominant influence on climate and the environment. There is ongoing debate as to when the Anthropocene would begin, but many suggest the 1950s since it would coincide with the presence in future rocks of an abundance of plastics and radioactive isotopes from nuclear weapons testing.

Blockchain – A digital ledger that can record transactions between parties in a reliable way, stored on several computers across a peer-to-peer network, both making it efficient and reducing the potential for error and corruption. It was initially developed to enable cryptocurrencies, like bitcoin, to operate efficiently. But the same technology can be used to trace supply chains, and hence can verify whether a product such as timber or tuna meat has come from a *sustainable* source.

Biodiversity (biological diversity) – A term that attempts to sum up the variety of life in the world. It is a function of the number of species, all the different kinds of animals, plants, fungi, and even micro-organisms like bacteria, and the number, or abundance, that exists of each of those species. In more abstract terms, the planet's biodiversity encapsulates not only millions of species and billions of individuals, but the trillions of different characteristics that those individuals have. The greater the biodiversity, the more the *biosphere* is able to deal with change, maintain balance and support life.

Biochar – A charcoal-like material that can be made from waste organic matter by baking it in a low- or zero-oxygen environment. It is under investigation as a viable approach to *carbon capture and storage*. It can be used as a building material or a *bioenergy* fuel, or to enrich soils and help them to retain water.

Bioenergy (biomass energy) – Renewable energy made available from materials derived from the living world. Fuels which are burned or digested for bioenergy include wood and fast-growing crops such as corn, soy, miscanthus and sugarcane. Biomass can be burned to generate electricity or converted into biofuel for transport fuels.

Conservancy – Simply an area that aims to protect the natural habitat but, in the context of this book, it refers to a protected area managed by the local community in a *sustainable* and economically viable manner.

Carbon budget (global) – The cumulative amount of carbon dioxide emissions estimated to limit global surface temperature to a certain level. Delay in cutting global emissions will use up the carbon budget faster and increases the risks of more global warming.

Carbon capture and storage (CCS) – The process of capturing carbon dioxide, usually from a large point-source such as a factory or power station, transporting it to an underground storage site, and depositing it for permanent storage so that it doesn't enter the atmosphere. CCS on a modern industrial site can reduce carbon dioxide emissions by up to 90 per cent, but increases operational energy use and costs. If combined with *bioenergy* generation (known as BECCS), or with direct air capture (DACCS) which scrubs carbon dioxide from ambient air, the CCS can theoretically remove carbon dioxide from the atmosphere, creating so-called 'negative emissions'.

These technologies, however, are in the research and development stage. *Nature-based solutions* offer a natural form of CCS (technically, carbon dioxide removal) that in addition increase *biodiversity.*

Carbon offset – A reduction in emissions of *greenhouse gases* aimed to compensate for, or balance, ongoing emissions elsewhere that cannot be avoided. Offsetting is done via the purchase of carbon credits or units which are measured in tonnes of carbon dioxide equivalent (CO_2e). Governments and large companies might choose to offset to comply with their obligations if it is cheaper than reducing domestically. Companies and individuals can purchase carbon offsets in a voluntary market to compensate for the emissions of their activities, for example, air travel – here the money spent on offsets typically funds development of *renewables, bioenergy* or *reforestation.* Offsetting should only be done as part of a broader emissions reduction strategy and in the long term is not a complete solution.

Carbon tax – A tax levied on the burning of carbon-based fuels (coal, oil, gas) to have polluters pay for the climate damage caused by the *greenhouse gas* emissions of their activities. It is proven to be an effective driver of emissions reductions in many sectors.

Carrying capacity – The maximum population size of a biological species that can be sustained in a specific environment, given the food, habitat, water, and other resources available.

Clean meat (cultured meat) – Meat for consumption produced as a cell culture of animal cells rather than from the slaughter of animals. It is a form of cellular agriculture. Research suggests that clean meat production has the potential to be much more efficient and environmentally friendly than traditional meat production, as it requires a fraction of the land, energy needs and water, and emits far fewer greenhouse gases per kilogram produced. It also has fewer animal welfare issues.

Circular economy (cyclical economy) – An economic system that aims to eliminate waste and the continual use of resources. Circular economies employ sharing, reuse, repair, refurbishment, remanufacturing and recycling to create a close-loop system. All waste becomes food for the next process, hence it is in contrast to the traditional linear economy, which has a take-make-use-discard model of production.

Culture – To a biologist, culture refers to a collection of behaviours, habits and skills that can be passed from one animal to another by non-genetic means, mainly imitation. In this sense, a culture is a parallel form of inheritance to biological (genetic) inheritance, and it undergoes its own form of evolution over time. Only a few species have been found to show evidence of culture, for example, chimpanzees, macaques and bottlenose dolphins. For humankind, cultural evolution is now the dominant form of evolution.

Demographic transition – A phenomenon occurring in nations in which there is a shift over time from high birth rates and high infant death rates in societies with minimal technology, education and economic development, to low birth rates and low death rates in societies with advanced technology, education and economic development.

Domestication – The process by which human beings assume a significant degree of influence over the reproduction and care of another species. Examples of plant domestications include wheat, potatoes and bananas. Examples of animal domestications include cattle, sheep and pigs. Domestication is the basis of all farming.

Doughnut Model – A reinterpretation of the *planetary boundaries* model, developed by Oxford economist Kate Raworth, that incorporates the basic needs of people as a social foundation, in addition to the existing ecological ceiling, and therefore defines a safe and just space for humanity. The idea is that we must keep below the ceiling, but not at the expense of the well-being of people. As such it acts as a framework for *sustainable* development.

Earth system – The integrated geological, chemical, physical and biological system of planet Earth. For the entire period of the *Holocene*, this system has maintained a benign environment for life, relying upon the complementary interaction of the atmosphere (air), hydrosphere (water), cryosphere (ice and *permafrost*), lithosphere (rock) and *biosphere* (life). The Earth system should continue to operate effectively and provide a benign environment as long as we keep within the *planetary boundaries*.

Ecological footprint – A measure of human impact on the environment. It essentially measures the quantity of nature it takes to support people or an economy and cope with our pollutants (especially greenhouse gases), and is expressed as a unit of area, the global hectare (gha). Currently we are demanding more global hectares than exist on Earth, hence the *Great Decline.*

Ecology – A branch of biology that studies the interactions and relationships between organisms and between organisms and their environment.

Forest dieback – The phenomenon of a stand of trees losing health and dying. Two of the major *tipping points* predicted to occur this century as a result of continuing deforestation and climate change are forest diebacks, one in the Amazon, the second in the boreal evergreen forest in Canada and Russia.

Forest transition – A pattern of land use change in an area over time as it is developed by a human society. To begin with, when the society is less developed, the forest is dominant. As the society develops and grows, expanding its food production, there is deforestation. As agriculture becomes more efficient and the population moves to urban areas, there can be *reforestation.* Several nations have been found to undergo a forest transition, and there are suggestions that we may also talk of a global forest transition involving the whole Earth.

Geoengineering (climate engineering) – The study and practice of forms of deliberate large-scale intervention in the *Earth system* in order to moderate and mitigate climate change. Some methods hope to boost the Earth's capacity to remove *greenhouse gases* from the environment, e.g., the fertilisation of the ocean with iron to raise phytoplankton productivity and increase carbon dioxide uptake in the surface waters. Other methods include solar radiation management, where, for example, aerosols are added to the stratosphere in the hope of reflecting more sunlight back out to space and thereby reducing global warming. Geoengineering is often criticised as untested and potentially very damaging to the environment and ourselves.

Great Acceleration – The dramatic, simultaneous surge in growth rate across a large range of measures of human activity, first recorded in the mid twentieth century and continuing to this day. The demand for resources and production of pollutants during the period of the Great Acceleration is the direct cause of much of the environmental degradation we see today.

Great Decline – The dramatic, simultaneous decline in a large range of environmental measures across the world, including biodiversity and climate stability, from the latter half of the twentieth century and continuing to this day. The decline is expected to escalate during this century, upon reaching a series of *tipping points*, and result in the radical destabilisation of the *Earth system*.

Green growth – A path of economic growth that uses resources in a *sustainable* manner. It is used to provide an alternative concept to traditional economic growth, which typically does not account for environmental damage.

Greenhouse gases (GHGs) – Gases that alter solar radiation and lead to the greenhouse effect which creates a 'blanket' that keeps the Earth at a higher ambient temperature. The primary greenhouse gases in Earth's atmosphere are water vapour, carbon dioxide, methane, nitrous oxide and ozone. Human activity has led to increased atmospheric concentration of some GHGs such as carbon dioxide, methane and nitrous oxide, which traps more heat and leads to climate change.

Gross Domestic Product (GDP) – A measure of productivity that summarises all the values of goods and services produced by a nation or sector over a given period. While it can be used as a measure of the productivity of a nation, GDP is widely criticised for not representing equality, well-being or environmental impact. Simon Kuznets, who developed GDP, warned that it should not be used as a measure of the welfare of a nation.

Holocene – A geological epoch, beginning about 11,700 years ago after the last glacial period. It has been a strikingly stable period of history, and corresponds with a rapid growth in humankind brought about by the invention of agriculture.

Hunter-gatherer – A culture in which a human society collects its food from the wild. It was the culture of all humans for 90 per cent of our history, until farming was invented at the start of the Holocene.

Hydroponics – A method of growing plants without soil by using a nutrient solution dissolved in water. It has various advantages, chiefly, hydroponics requires much less water to grow plants.

Lag phase – An initial phase on a growth curve in which there is little net growth due to some restricting factor or factors.

Log phase – A phase on a growth curve characterised by logarithmic or exponential growth.

Marine Protected Areas (MPAs) – Protected areas of seas or ocean that restrict human activity to some degree, such as limiting fishing practices, seasons or catches. No-fish or no-take zones ban fishing of all kinds entirely. Currently there are over 17,000 MPAs worldwide, representing just over 7 per cent of the ocean.

Mass extinction – A widespread and rapid decline in the *biodiversity* of Earth. According to most authorities, a mass extinction event has occurred at least five times in life's history including that which brought an end to the dinosaurs.

Micro-grid – A micro-grid is a localised group of electricity sources that can operate in association with or remote from a regional grid. Because they work together to supply electricity, they cope better with surges in demand than solo generators. They are becoming more common now that distributed generation of electricity using *renewables* is becoming more affordable.

Nature-based solution – The use of nature to jointly tackle social and environmental issues, especially climate change, water security, food security, pollution and disaster risk. Examples include planting mangroves to prevent coastal erosion, *MPAs* to increasing fishing catch, greening cities to reduce air temperature, building wetlands to prevent flooding, and *reforestation* to act as a natural *carbon capture and storage* facility. Nature-based solutions are often relatively cost-effective and have the significant benefit of increasing *biodiversity*.

Ocean acidification – The ongoing decrease in the pH of the ocean caused by the uptake of carbon dioxide from the atmosphere. Seawater is slightly alkaline, so ocean acidification initially refers to a move to neutral conditions. As it continues, the acidification damages much of the life in the ocean. When it has occurred previously in Earth's history it has been accompanied by a *mass extinction event* and a long-lasting decline in the efficiency of the *Earth system*.

Ocean forestry – A proposal for a *nature-based solution* to climate change in which seaweed forests are grown and farmed. As they grow they act as a *carbon capture and storage* system, and the seaweed produced can be used for *bioenergy*, food or permanently disposed of to remove the carbon from the atmosphere.

Overfishing – The removal of a species of fish from a body of water at a rate that the species cannot replenish, resulting in those species becoming underpopulated in that area. In 2020, the Food and Agriculture Organization of the United Nations reported that one-third of world fish stocks are overfished.

Peak catch – The point in time at which the weight of fish landed stops increasing. We reached peak catch in the mid 1990s. After that point there has been a slight decline in global catch.

Peak child – The point at which the number of children (commonly regarded as those under 15) globally stops increasing. The UN currently predicts that peak child will happen towards the middle of the century.

Peak farm – The point at which the area devoted to farmland stops increasing. The UN's Food and Agriculture Organization predicts that this will happen by about 2040.

Peak human – The point at which the human population stops increasing. The UN's Population Division currently predicts that peak human will happen in the early twenty-second century at about 11 billion people. However, by lifting people out of poverty and empowering women, it is forecast we could reach peak human as soon as 2060 at just 8.9 billion people.

Peak oil – The point in time at which global production of oil is at its maximum, after which oil extraction will decline.

Permafrost – Ground, often below the surface, that remains continually frozen. Permafrost on land is most extensive in the tundra and Arctic regions of Russia, Canada, Alaska and Greenland. As the globe warms, the permafrost is predicted to thaw, releasing methane, a powerful *greenhouse* gas, into the atmosphere, thus entering a positive feedback loop in which more permafrost then thaws, leading to a *tipping point* and runaway global warming.

Perpetual growth – The assumption that underpins our current economic model, that *Gross Domestic Product* will continue to increase, year on year, forever. In reality, many developed economies have very low increases in GDP each year, between 0 and 2 per cent, but that is, of course, still growth.

Phytoplankton – The photosynthesising organisms in the microscopic but widespread plankton community living in the surface waters of the ocean. Phytoplankton are the basis of many marine food chains.

Planetary boundaries – A concept developed by *Earth system* scientists Johan Rockström and Will Steffen in order to define a safe operating space for humanity. The team used data from multiple sources to define nine factors that influence the stability of the Earth system. They calculated the degree to which current human activity is impacting upon those factors and established thresholds, that, if crossed, may lead to potentially catastrophic change. The nine factors are: *biodiversity* loss, climate change, chemical pollution, ozone depletion, atmospheric aerosols, *ocean acidification*, nitrogen and phosphorus use, freshwater consumption and land-use change (from wild space to fields or plantations). Of these nine, the team have identified two – climate change and *biodiversity* loss – as the 'core boundaries' in that they are both affected by all the other boundaries and could alone, if crossed, bring about the destabilisation of the planet. They advise that currently, humankind has crossed four boundaries: climate change, biodiversity loss, land-use change and the use of nitrogen and phosphorus. They therefore report that the Earth system is already in an unstable state.

Plant-based diet – A diet that consists mostly or entirely of foods from plants, with few or no animal products. A plant-based diet is more *sustainable* than contemporary diets containing many animal products since it, on average, takes up less land, energy and water to produce, and leads to the emission of fewer *greenhouse gases.*

REDD+ – A UN initiative that stands for 'Reducing Emissions from Deforestation and forest Degradation and the role of conservation, *sustainable* management of forests and enhancement of forest carbon stocks in developing countries'. REDD+ attempts to create financial value for carbon stored in standing forests, creating more incentives for retaining the forest with the ambition of reducing deforestation and forest degradation in developing countries.

Regenerative farming – A conservation and rehabilitation approach to agriculture, focusing on increasing the natural health of soils. It is a reaction against industrial farming which typically decreases soil health over time and requires supplements of fertilisers and pesticides. Regenerative farming techniques lead to soils with increased organic content, *carbon capture and storage* capability and soil *biodiversity.*

Reforestation – The natural or intentional return of native forests and woodlands. Reforestation can be used as a blanket term, or specifically to areas that have been recently deforested. In this case, afforestation applies to areas that have not been forested for some time, e.g. traditional farmland, or within cities. Reforestation is a potential *nature-based solution* to climate change for it can lead to significant *carbon capture and storage*.

Renewables (renewable energy) – Energy from sources that naturally replenish themselves on a human timescale such as solar, wind, bioenergy, tidal, wave power, hydroelectric power and geothermal heat. Renewables are typically lower- or zero-carbon replacements for fossil fuels.

Rewild – The process of restoring and expanding biodiverse spaces, communities and systems. Rewilding is often large-scale, seeking to reinstate natural processes and, where appropriate, missing species. In some cases proxy species may be used to perform a similar role to missing species within the recovering community. In this book, the term rewild is used in its broadest sense, meaning the ambition to restore nature across the Earth, reversing *biodiversity* loss by ensuring that humankind as a whole becomes more *sustainable*. Thus climate change mitigation is regarded as a necessary component of rewilding the world.

Shifting baseline syndrome – The tendency for the concept of what is 'normal' or 'natural' to change over time due to the experiences of subsequent generations. In this book, it is a term used to describe our own capacity to forget, over the generations, how *biodiverse* a natural environment should be.

Silvopasture – One of a number of regenerative agriculture techniques, silvopasture is the practice of raising domesticated animals alongside trees or within woodlands and forests. It can significantly increase the health and yield of the animals, since they are sheltered by the trees and are able to browse as well as graze.

Spill-over effect – The phenomenon of improvements in the *biodiversity* of one area benefitting the biodiversity of neighbouring areas. Specifically, a spill-over effect is experienced in the waters surrounding *MPAs*, in which fish stocks recovering in the MPA spill over into the neighbouring areas, increasing fishing catch.

Sustainable (sustainability) – Literally, the ability for something to continue forever. In the context of this book, it refers to the capacity for humankind and the *biosphere* to coexist permanently. To be sustainable, humankind must establish a life on our planet that exists within the *planetary boundaries*.

Sustainability revolution – A predicted, coming industrial revolution in which the driver is a wave of innovation focused upon sustainability. It will feature *renewables*, low-impact transport, a zero-waste *circular economy, carbon capture and storage, nature-based solutions, alt-proteins, clean meat, regenerative agriculture, vertical farming,* etc. It promises an opportunity for *green growth* and an aspirational future.

Tipping point – A threshold that, when exceeded, can lead to an abrupt, large, often self-amplifying and potentially, irreversible change in the *Earth system*.

Trophic cascade – An effect in an ecosystem in which change in one level in a food chain, known as a 'trophic level', triggers multiple knock-on effects within others. In history, as we removed top predators, there will have been trophic cascades that radically altered ecosystems and hence whole landscapes and seascapes. For example, in removing wolves, deer populations are able to increase, preventing natural *reforestation*. By returning top predators as we *rewild*, we can bring about trophic cascades that reinstate natural *biodiversity*, as demonstrated by the reintroduction of the wolf in Yellowstone National Park.

Urban farming – The production of food and other agricultural products in and around urban areas. Urban farming is often highly *sustainable* in that it uses land already occupied by humankind, reduces transportation and employs methods such as *hydroponics* and *renewables* to produce food.

Vertical farming – The practice of producing food in vertically stacked layers often in a controlled environment and using *hydroponics* or *aquaponics*. It is often a highly *sustainable* approach to farming certain types of plants in that it produces more food from less land and can operate without fertilisers or pesticides.

Wildland farm – A *rewilding* approach to farming in which a community of different livestock animals which mimics the natural community of the location is permitted to wander freely about a farm without supplements. The animals are kept in numbers that suit the *carrying capacity* of the landscape, and bring about a *trophic cascade* in which the *biodiversity* of the land increases.

Acknowledgements

A Life on Our Planet as a project, comprising both this book and an accompanying film, has been several years in the making and required the assistance and contributions of many colleagues. The idea initially came about during conversations with Colin Butfield at the Worldwide Fund for Nature (WWF), and with Alastair Fothergill and Keith Scholey, my old friends at Silverback Films. I am indebted to all three of them. They were instrumental in defining the structure of this book and they led the production of the film, which informed so much of its contents.

My primary debt, however, in writing the book has been to my co-author, Jonnie Hughes. He has been involved with environmental issues for many years and was a director on the film. His eloquence, expertise and clarity of thought have been invaluable. This has been particularly so in the third part of this book which draws upon the ideas, opinions and research of people from many fields and organisations.

We could not have hoped to compile such a vision without the substantial assistance of the Science Team at WWF. We would like to thank in particular, Mike Barrett, Executive Director of Conservation and Science at WWF-UK, for sharing his clear perspective on the environmental crisis, and for guiding the team that works with him on their milestone publication, the *Living Planet Report*, which has been of such inspiration to all of us involved with this project. Our thanks also go to Mark Wright, Science Director at WWF, who has put in many long hours, ensuring that the arguments presented across the whole project are rooted in real-world examples and good scientific research.

This collaboration with WWF introduced us to many inspiring communicators and researchers too numerous to list here. We would, however, like to especially acknowledge Johan Rockström and the team that worked with him when creating the planetary boundaries model, and Kate Raworth, author of the Doughnut Model. Their work has brought profound insights at this critical moment in our history. The writings and research of Paul Hawken and Callum Roberts have been instrumental in understanding the problems and solutions associated respectively with climate change and the ocean.

We are both very grateful for the guidance of Albert DePetrillo and Nell Warner at Penguin Random House, and to Robert Kirby and Michael Ridley for their assistance in the production of this book.

My thanks also go to my dear daughter, Susan, who organises me and my diary and has listened with extraordinary patience to every word of this book – several times.

ACKNOWLEDGEMENTS

Engaging in this project has brought many emotions. The truth of our planet's current predicament is beyond alarming. Learning the very latest detail of our crisis has troubled me greatly. But, counter to that, it is heart-warming to discover the extent to which brilliant minds are now at work to understand and, further, to solve the problems we face. My great hope is that these minds may soon come together in a position to influence our future. As I have been reminded during the creation of *A Life on Our Planet*, it is possible to achieve so much more working with others than any one of us can achieve alone.

David Attenborough
Richmond, UK
8 July 2020

Picture Credits

Notes

Part One: My Witness Statement

1   The most reliable source of world population data is compiled by the United Nation's Population Division, and a broad range of information can be accessed via https://population.un.org/wpp/ and in particular the 'World Population Prospects 2019 – Highlights' at https://population.un.org/wpp/Publications/Files/WPP2019_Highlights.pdf.

2   Here we are using 'carbon' as a shorthand for 'carbon dioxide'. The increasing proportion of carbon dioxide in the atmosphere is a characteristic of our recent development and a big driver of global warming. Accumulation in the atmosphere is directly linked to the burning of fossil fuels – coal, oil and gas. Throughout this book, we use carbon dioxide data from the Mauna Loa observatory: https://www.esrl.noaa.gov/gmd/ccgg/trends/data.html.

3   Estimates on remaining wilderness are based on data and extrapolations from Ellis E. et al (2010), 'Anthropogenic transformation of the biomes, 1700 to 2000 (supplementary info Appendix 5)', *Global Ecology and Biogeography* 19, 589–606.

4   The exact number of mass extinction events depends upon at what point you determine a large extinction event to be 'mass'. Typically, geologists talk of five mass extinction events before present, in order, the Ordovician-Silurian event of 450 million years ago (Ma), the Late Devonian event (375 Ma), the Permian-Triassic event (252 Ma), which was the most extreme extinction event with up to 96 per cent of marine and 70 per cent of terrestrial species disappearing, the Triassic-Jurassic event (201 Ma) and the Cretaceous-Paleogene event (66Ma) which ended the age of the dinosaurs.

5   There are a number of theories concerning what brought about the end of the age of the dinosaurs. The proposal that it was largely due to the impact of a meteorite on the Yucatan Peninsula was viewed as radical when first suggested but, with increasing evidence, including, most recently, deep-rock drilling in the Chicxulub crater in 2016, it has become the most widely supported theory. For a good recent account of this evidence, see Hand, E. (2016), 'Drilling of dinosaur-killing impact crater explains buried circular hills', *Science*, 17 November 2016, https://www.

sciencemag.org/news/2016/11/updated-drilling-dinosaur-killing-impact-crater-explains-buried-circular-hills.

6   Genetic analysis supports the belief there was a population bottleneck approximately 70,000 years ago, where humanity's numbers dropped to a very low level. There is a vigorous debate about what caused this specific bottleneck – ranging from a volcano to sociocultural reasons – but most believe the underlying reason our population wasn't large enough to easily weather any such events was the long-term unpredictability of the climate. For the interested reader, these are just a few of the articles that explore the bottleneck: Tierney J.E. et al (2017) 'A climatic context for the out-of-Africa migration'_https://pubs.geoscienceworld.org/gsa/geology/article/45/11/1023/516677/A-climatic-context-for-the-out-of-Africa-migration'; Huff, C.D. et al (2010), 'Mobile elements reveal small population size in the ancient ancestors of *Homo sapiens*', https://www.pnas.org/content/107/5/2147; Zeng, T.C. et al (2018), 'Cultural hitchhiking and competition between patrilineal kin groups explain the post-Neolithic Y-chromosome bottleneck', *Nature*, https://www.nature.com/articles/s41467-018-04375-6.

7   We can judge the average temperature of past environments by examining ice cores, tree rings and ocean sediments. This tells us that, for several hundred thousand years before the Holocene, the average temperature of the Earth was far more erratic and generally cooler than today's average. NASA have produced an interesting article that gives more information: https://earthobservatory.nasa.gov/features/GlobalWarming/page3.php.

8   The logs of all the communications of the Apollo missions are available via the NASA website, and make fascinating reading: https://www.nasa.gov/mission_pages/apollo/missions/index.html.

9   The important role of whales in distributing nutrients is just now coming to light. Whales transport nutrients laterally, in moving between feeding and breeding areas, and vertically, by transporting nutrients from nutrient-rich deep waters to surface waters via faecal plumes and urine. It is estimated that the capacity of animals to move nutrients away from patches where it is concentrated has decreased to about 5 per cent of what it was before industrial whaling. See Doughty, C.E. (2016), 'Global nutrient transport in a world of giants' https://www.ncbi.nlm.nih.gov/pmc/articles/PMC4743783/. For a localised study in the Gulf of Maine, see Roman, J. and McCarthy, J.J. (2010), 'The Whale Pump: Marine Mammals Enhance Primary Productivity in a Coastal Basin', PLoS ONE 5(10): e13255, https://doi.org/10.1371/journal.pone.0013255.

10   The first global estimate of the impact of whaling was completed only recently; it revealed that whaling may have been the largest global cull of any animal by weight in human history. See Cressey, D. (2015), 'World's whaling slaughter tallied', *Nature*, https://www.nature.com/news/world-s-whaling-slaughter-tallied-1.17080.

11   The website www.globalforestwatch.org is a useful resource online that aims to chart all change in global forest cover. There are difficulties in doing so. Plantations can appear to be natural forest from space, whereas they are in fact very low-diversity habitats in comparison. The Global Forest Biodiversity Initiative https://www.gfbinitiative.org/ is attempting to more accurately chart the biodiversity of forests. One of its lead members, Thomas Crowther, recently assessed the global tree total and estimated its depletion at our hand. See 'Mapping tree density at a global scale', *Nature* 525, 201–205 (2015), https://doi.org/10.1038/nature14967.

12   In 2016, the IUCN estimated that the Borneo orangutan numbered 104,700 individuals. This represents a decline from an estimated 288,500 individuals in 1973. They predict a further decline of 47,000 individuals by 2025; https://www.iucnredlist.org/species/17975/123809220#population.

13   Eukaryotic cells are widely estimated to have evolved between 2 and 2.7 billion years ago, so roughly 1.5 billion years after the origin of life; https://www.scientificamerican.com/article/when-did-eukaryotic-cells/. Multicellular life evolved just over half a billion years ago, roughly 1.5 billion years later; https://astrobiology.nasa.gov/news/how-did-multicellular-life-evolve/.

14   A study of the world's fishing catch data was conducted by researchers in 2003 and revealed the startling rate at which our fishing effort reduced the largest fish in the sea. See Rupert Murray's film *The End of the Line* for an interview on this research, or the paper, Myers, R. and Worm, B. (2003), 'Rapid Worldwide Depletion of Predatory Fish Communities', *Nature* 423, 280–3, https://www.nature.com/articles/nature01610.

15   For an up-to-date assessment of the impact of fishing subsidies worldwide, see Sumaila et al (2019), 'Updated estimates and analysis of global fisheries subsidies', https://doi.org/10.1016/j.marpol.2019.103695; WWF (2019), 'Five ways harmful fisheries subsidies impact coastal communities', https://www.worldwildlife.org/stories/5-ways-harmful-fisheries-subsidies-impact-coastal-communities.

16   For more of these historical stories, and a detailed description of the ways in which shifting baseline syndrome has impacted on the

expectations we have for our ocean, see Callum Roberts (2013), *Ocean of Life*, Penguin Books.

17   A thorough appraisal of the extinction at the end of the Permian is here: White, R.V. (2002), 'Earth's biggest "whodunit": unravelling the clues in the case of the end-Permian mass extinction', *Philosophical Transactions of the Royal Society of London* 360 (1801): 2963–2985. Available at https://www.le.ac.uk/gl/ads/SiberianTraps/Documents/White2002-P-Tr-whodunit.pdf.

18   The situation in the Arctic and Antarctic is rapidly changing year on year. For the best source of the latest data, both these sites are very interesting and authoritative: National Snow and Ice Data Center, https://nsidc.org/data/seaice_index/ and National Oceanic and Atmospheric Administration, https://www.arctic.noaa.gov/Report-Card. For more detail, the World Glacier Monitoring Service (WGMS) also collects yearly data of all the world's monitored glaciers (https://wgms.ch/).

19   The most comprehensive report on the state of world biodiversity is the IPBES Global Assessment (2019). The summary report is available at https://ipbes.net/sites/default/files/2020-02/ipbes_global_assessment_report_summary_for_policymakers_en.pdf. In addition, the WWF's biannual *Living Planet Report* offers an authoritative and highly accessible stocktake; visit www.panda.org for the latest edition.

20   The UN's Food and Agriculture Organisation (FAO) publishes the most comprehensive review on the marine and freshwater fish sector every two years, entitled *The State of World Fisheries and Aquaculture*. Find the 2020 edition here: http://www.fao.org/state-of-fisheries-aquaculture.

21   Riskier Business (2020) gives a detailed account of how much land is required, outside of the UK, to supply UK demand for just seven commodities (including soy and beef). A summary and the full report can be downloaded from https://www.wwf.org.uk/riskybusiness.

22   An accessible review of global insect loss is Goulson, D. (2019), 'Insect declines and why they matter'; it can be found at https://www.somersetwildlife.org/sites/default/files/2019-11/FULL%20AFI%20REPORT%20WEB1_1.pdf. And for those who want to read about restoring insect populations, some good examples (from the UK) can be found in Wildlife Trusts (2020), 'Reversing the decline of insects', https://www.wildlifetrusts.org/sites/default/files/2020-07/Reversing%20the%20Decline%20of%20Insects%20FINAL%2029.06.20.pdf. See also Chapter 2, note 9.

23   These figures for the representation of different groups come from a groundbreaking assessment of life on Earth, Bar-On, Y.M., Phillips, R.

and Milo, R. (2018), 'The biomass distribution on Earth', *Proceedings of the National Academy of Sciences* 115 (25) 6506–6511, https://www.pnas.org/content/pnas/early/2018/05/15/1711842115.full.pdf.

Part Two: What Lies Ahead

1   Two leading bodies are dedicated to reporting on the state of the planet. The Intergovernmental Panel on Climate Change (IPCC) is the best source of information on the consensus of current and forecast climate change (www.ipcc.ch). The Intergovernmental Platform on Biodiversity and Ecosystem Services (IPBES) is the best source of information on the state of biodiversity (www.ipbes.net). For those interested in the concept of tipping points, a helpful review is McSweeney, R. (2010), 'Explainer: Nine "tipping points" that could be triggered by climate change', available at https://www.carbonbrief.org/explainer-nine-tipping-points-that-could-be-triggered-by-climate-change.

2   For a detailed account of this work and its implications, the very readable Rockström, J. and Klum, M. (2015), *Big World, Small Planet*, Yale University Press, is recommended.

3   The latest study by the IPBES (2019) suggests that the current rate of extinctions is tens to hundreds of times the rate of the average over the last 10 million years, and the average rate of vertebrate species loss over the last century is thought to be up to 114 times higher than the background rate. See https://ipbes.net/global-assessment.

4   Among those who predict an Amazon dieback in the near-term is Brazilian Earth system scientist Carlos Nobre. An informative interview with Nobre can be found here: https://e360.yale.edu/features/will-deforestation-and-warming-push-the-amazon-to-a-tipping-point. A corresponding paper is here: Nobre, C.A. et al (2016), 'Land-use and climate change risks in the Amazon and the need of a novel sustainable development paradigm', https://www.pnas.org/content/pnas/113/39/10759.full.pdf.

5   The best sources for the latest figures of ice loss are the IPCC *Special Report on the Ocean and Cryosphere in a Changing Climate* (2019), https://www.ipcc.ch/srocc/, and the *Arctic Monitoring and Assessment Programme Climate Change Update 2019: An Update to Key Findings of Snow, Water, Ice and Permafrost in the Arctic (SWIPA) 2017*, https://www.amap.no/documents/doc/amap-climate-change-update-2019/1761.

6   For information relating to the permafrost, the Global Terrestrial Network for Permafrost (https://gtnp.arcticportal.org/) includes all the recent data.

7   A key source of data on bleaching events and coral reef loss is the US government's NOAA Coral Reef Watch, https://coralreefwatch.noaa.gov, which uses satellite data together with geographical information systems to monitor sea conditions across the world. For more detail, I'd also recommend the Global Coral Reef Monitoring Network reports: https://gcrmn.net/products/reports/.

8   The UN's Food and Agriculture Organization produces frequent reports on the state of global agriculture and food production. One of its keystone reports is its *Status of the World's Soil Resources* from 2015, which laid out the chief concerns over the sustainability of modern, industrial agriculture: http://www.fao.org/3/a-i5199e.pdf.

9   A worldwide decline in insect life is widely acknowledged. Forecasts for insect biodiversity loss in the future are harder to assess, but a leading and well-respected paper was completed by Francisco Sanchez-Bayo and Kris Wyckhuys in 2019; see 'Worldwide decline of the entomofauna: A review of its drivers', https://www.sciencedirect.com/science/article/pii/S0006320718313636. See also Chapter 1, note 22.

10   During the COVID-19 pandemic, the IPBES (2020) made clear the link between emergent viruses and our degradation of the environment in a guest article; see https://ipbes.net/covid19stimulus.

11   The IPCC is the leading international body assessing the science of climate change. Its 2019 report on *Oceans and the Cryosphere in a Changing Climate* includes the latest projections of sea level rise: https://www.ipcc.ch/srocc/chapter/summary-for-policymakers/.

12   The C40 cities organisation is a network of the world's megacities committed to addressing climate change. It is a good source of information on how urban areas are likely to be affected by global warming, and how responsible cities are tackling the issues they face. See https://www.c40.org.

13   There are many models that project future impacts of climate change. The modelling that our planet may be 4°C warmer by 2100 comes under scenario RCP8 of the IPCCC 5th assessment, https://www.ipcc.ch/assessment-report/ar5/. The projection that one-quarter of the human population could live in places with an average temperature over 29°C uses a different set of modelling assumptions that, whilst based on the more extreme end of projections, is still considered a possible outcome. See Xu, C. et al (2020), 'Future of the human climate niche', *Proceedings of*

the National Academy of Sciences May 2020, 117 (21), 11350–11355, https://www.pnas.org/content/early/2020/04/28/1910114117.

Part Three: A Vision for the Future: How to Rewild the World

1   This comes from *The Dasgupta Review: Independent Review on the Economics of Biodiversity*, due out in late 2020. This review will present a powerful argument for valuing the environmental services of nature more appropriately within a modern economy. See https://www.gov.uk/government/publications/interim-report-the-dasgupta-review-independent-review-on-the-economics-of-biodiversity.

2   Kate Raworth's book *Doughnut Economics* (2017) is an excellent appraisal of the incompatibility of our current economic system with the realities of the natural world. It contains a detailed description of the Doughnut Model and offers much guidance on how we may organise our economies sustainably.

3   Tropical rainforests are in many cases ancient ecosystems. A good overview of their history and how they function can be found in Ghazoul, J. and Sheil, D. (2010), *Tropical Rain Forest Ecology, Diversity, and Conservation*, Oxford University Press.

4   *The Dasgupta Review: Independent Review on the Economics of Biodiversity – an interim report* proposes that, as an alternative to using GDP to assess success, we should turn to a Net Domestic Product (NDP) measure that includes the true cost of environmental damage; see https://www.gov.uk/government/publications/interim-report-the-dasgupta-review-independent-review-on-the-economics-of-biodiversity. For more information on the Happy Planet Index, see http://happyplanetindex.org/.

5   The primary source of this data and a good source for global energy information is the International Energy Agency (www.iea.org).

6   The world of carbon budgets is a very technical area. For an overview see https://www.ipcc.ch/sr15/chapter/chapter-2/. For an account of future emission projections, see https://ourworldindata.org/co2-and-other-greenhouse-gas-emissions#future-emissions.

7   Project Drawdown is a non-profit organisation that has compiled an extensive and highly readable analysis of measures to mitigate climate change, each one assessed for its relative significance; see www.drawdown.org.

8   For a radical forecast of the changes that may come to the transport industry, see https://www.rethinkx.com/transportation.

9   The Stockholm Resilience Centre is a guiding light in Earth system
    science and thinking on sustainability. It was behind the planetary
    boundaries model and works to advise governments on environmental
    policy. See more at https://www.stockholmresilience.org/.

10 For some of the best ways to deliver the energy transition, see the
    several WWF reports available at https://www.wwf.org.uk/updates/
    uk-investment-strategy-building-back-resilient-and-sustainable-economy.

11 Examples of studies that link greater biodiversity with a greater capacity
    to capture and store carbon in ecosystems include Atwood et al (2015),
    which demonstrates that, when top predators were removed, carbon
    capture and storage in saltmarshes in New England and in mangrove
    and seagrass ecosystems in Australia was reduced due to the rise of
    herbivores, https://www.nature.com/articles/nclimate2763; Liu et al
    (2018) found that tree species richness in subtropical rainforests in China
    increased the capacity of the forest to capture and store carbon, https://
    royalsocietypublishing.org/doi/full/10.1098/rspb.2018.1240; and Osuri et
    al (2020) found that natural forests were better at capturing and holding
    on to carbon than plantations in India, https://iopscience.iop.org/
    article/10.1088/1748-9326/ab5f75.

12 Useful information on the status of Marine Protected Areas is to be
    found at Protected Planet: https://www.protectedplanet.net/marine. It
    is important to note that at present not all protected areas are effectively
    managed. Indeed some estimates suggest only about 50 per cent of them
    are true, effectively run MPAs.

13 The Smithsonian has a detailed report on the Cabo Pulmo MPA success
    story which demonstrates how important it is to get the local community
    invested in MPAs and conservation projects in general; see https://ocean.
    si.edu/conservation/solutions-success-stories/cabo-pulmo-protected-area.

14 For more on the effectiveness of coastal ecosystems in capturing and
    removing carbon, and the efforts under way to restore mangroves,
    saltmarshes and seagrass meadows for this purpose, see https://www.
    thebluecarboninitiative.org/. To see more detail on design of Marine
    Protected Areas, this is an interesting read from Australia: https://ecology.
    uq.edu.au/filething/get/39100/Scientific_Principles_MPAs_c6.pdf.

15 The marine environment poses particular difficulties in assessing
    populations of fish stocks and monitoring the activities of fishing vessels
    at sea, both of which are needed to ensure sustainability. These issues are
    being grappled with by existing certification schemes but are not yet fully
    resolved.

16 The UN's Convention on the Law of the Sea is the presiding international treaty on the world's use of the ocean. It is currently being amended for the first time in decades, and many people are working hard to ensure that sustainability is at the heart of its refreshed contents. If we get these changes right, it could transform humankind's relationship with the ocean. For more information, see https://www.un.org/bbnj/.

17 Figures on both fishing catch and aquaculture production are reported regularly by the UN's FAO in their *State of World Fisheries and Aquaculture*. The 2020 edition can be found here: http://www.fao.org/state-of-fisheries-aquaculture.

18 The Aquaculture Stewardship Council (ASC) manages a certification and labelling programme for responsible aquaculture. Look for its green label on aquaculture products such as farmed salmon and shellfish. See https://www.asc-aqua.org/.

19 The technology of *Bioenergy with Carbon Capture and Storage* (BECCS) is currently under investigation as a method of removing carbon from the atmosphere whilst generating heat or electricity. If it proves to be a scalable option, it could help reduce the pressure of bioenergy crops that compete for space with food production or natural habitats. The advantage of using kelp as a bioenergy crop is that a restored kelp forest is a high-biodiversity habitat that grows at such speed it can withstand regular but well-managed harvesting.

20 For a vivid account of the ways in which humankind uses land, see the presentation created by the research and data project, Our World in Data: https://ourworldindata.org/land-use.

21 The IPCC *Special Report on Climate Change and Land* (revised 2020) has some fascinating insights on how land use impacts the climate: https://www.ipcc.ch/srccl/chapter/summary-for-policymakers/.

22 We are still to learn so much about the ways in which soils function. The micro-organisms and invertebrates that live in healthy soils interact with each other and with the plant life about them in numerous, complex ways. It is becoming apparent that high soil biodiversity is of fundamental importance to the fixing of key nutrients, the condition of the soil, the growth of plants and the capture and storage of carbon on land. See Hirsch, P. R. (2018), 'Soil microorganisms: role in soil health', in Reicosky, D. (ed.), *Managing Soil Health for Sustainable Agriculture*, Volume 1: 'Fundamentals', Burleigh Dodds, Cambridge, UK, pp. 169–96. For those who are looking for a good overview of the food production system and what needs to change, the following report from the Food

and Land Use Coalition 'demonstrates how, by 2030, food and land use
systems can help bring climate change under control, safeguard biological
diversity, ensure healthier diets for all, drastically improve food security
and create more inclusive rural economies': FOLU (2019), *Growing Better:
Ten Critical Transitions to Transform Food and Land Use*, available at: https://
www.foodandlandusecoalition.org/wp-content/uploads/2019/09/FOLU-
GrowingBetter-GlobalReport.pdf.

23 Wageningen University in the Netherlands is a leading research centre
investigating high-tech approaches to improving the sustainability of
agriculture and has been instrumental in many of the techniques being
trialled in some of these Dutch farms. See https://weblog.wur.eu/spotlight/.

24 Two leading sources of information on regenerative agriculture are
Regeneration International (https://regenerationinternational.org) and
Burgess, P.J., Harris, J., Graves, A.R., Deeks. L.K. (2019), *Regenerative
Agriculture: Identifying the Impact; Enabling the Potential*, Report for
SYSTEMIQ, 17 May 2019, Cranfield University, Bedfordshire, UK,
https://www.foodandlandusecoalition.org/wp-content/uploads/2019/09/
Regenerative-Agriculture-final.pdf.

25 For a presentation on how much of the world's land we would need
in order to feed the world population with the average diet of a given
country, see https://ourworldindata.org/agricultural-land-by-global-
diets. Data on meat consumption around the world can be found at
https://ourworldindata.org/meat-production#which-countries-eat-the-
most-meat.

26 The leading reports in recent times have been *The Planetary Health Diet
and You* by the EAT-Lancet commission (2019), see https://eatforum.
org/eat-lancet-commission/the-planetary-health-diet-and-you/, and the
FAO's *Sustainable Diets and Biodiversity* review (2010), see http://www.fao.
org/3/a-i3004e.pdf.

27 This assessment comes from a recent paper from the Programme on the
Future of Food at the University of Oxford; see Springmann, M. et al
(2016), *Analysis and valuation of the health and climate change cobenefits of
dietary change*, https://www.pnas.org/content/early/2016/03/16/
1523119113.

28 Original sources are quoted in https://www.theguardian.com/
business/2018/nov/01/third-of-britons-have-stopped-or-reduced-meat-
eating-vegan-vegetarian-report and https://www.foodnavigator-usa.com/
Article/2018/06/20/Innovative-plant-based-food-options-outperform-
traditional-staples-Nielsen-finds. A recent survey showed the number of

meat reducers in the UK has risen from 28 per cent in 2017 to 39 per cent in 2019; see https://www.mintel.com/press-centre/food-and-drink/plant-based-push-uk-sales-of-meat-free-foods-shoot-up-40-between-2014-19.

29 For a radical review of how quickly and extensively the agriculture sector may be changed by this revolution in food production, see https://www.rethinkx.com/food-and-agriculture-executive-summary. The FAO study (2012) looking at World Agriculture towards 2030/2050 is a very good detailed analysis; see http://www.fao.org/3/a-ap106e.pdf.

30 The amount of land that each human being needs to feed themselves with plant-based foods is actually decreasing rapidly due to the rising yields of modern farming. To see data on this trend and a series of predictions of the amount of farmland needed in the future, based on FAO data, see https://ourworldindata.org/land-use#peak-farmland.

31 More information on the UN's REDD+ programme can be found at https://www.un-redd.org/.

32 The Forestry Stewardship Council (FSC) is an international non-profit organisation whose mission is to promote environmentally appropriate, socially beneficial and economically viable management of the world's forests. It runs a global forest certification system. Its green label is a good indication that a timber or wood product has come from sustainably and equitably managed forests. See https://www.fsc.org.

33 A good example of sustainable tropical forestry is the Deramakot Forest Reserve in Sabah, Borneo, which has been certified as sustainable by the Forestry Stewardship Council since 1997, longer than any other tropical forest. Logging is carefully managed to retain biodiversity and, indeed, surveys have shown that the reserve has very similar biodiversity to untouched forest elsewhere in Sabah. An interesting story and short film about Deramakot can be found at https://www.weforum.org/agenda/2019/09/jungle-gardener-borneo-logging-sustainably-wwf/.

34 For example, in the UK the government is considering awarding subsidies to farmers on the basis of their land's 'public goods', including levels of biodiversity and carbon capture, rather than simply for cultivating the land as now. There are those who doubt that the policy will go far enough, but a recent survey by the Wildlife and Countryside Link has shown that the farming community in England, at least, supports this transition. See https://www.wcl.org.uk/assets/uploads/files/WCL_Farmer_Survey_Report_Jun19FINAL.pdf.

35 The story of Charlie and Isabella's rewilding of their farm in Sussex is wonderfully captured in the book *Wilding* by Isabella Tree (2018). It is

a revelatory account of both the issues with our modern approach to farming and the startling degree to which nature can recover if given the chance. It also demonstrates the environmental services that we gain from a diverse ecosystem. The farm has become substantially better equipped at capturing carbon, improving the health of its soils and mitigating flooding.

36 Rewilding projects are gaining a foothold all over the world and are increasingly being adopted as an approach that enables the recovery of biodiversity and natural processes at a landscape scale. Examples include: the Ennerdale project in a mixed-use production landscape in the heart of one of the UKs most well-loved locations, the Lake District; the American Prairie Reserve initiative in the USA, linking and restoring native prairie grasslands; and projects across Europe supported by Rewilding Europe such as the restoration of the Danube delta. For more information, see http://www.wildennerdale.co.uk/, https://rewildingeurope.com/space-for-wild-nature/ and https://rewildingeurope.com/areas/danube-delta/.

37 For Yellowstone National Park's own account of the wolf recovery and its effect on biodiversity, see https://www.nps.gov/yell/learn/nature/wolf-restoration.htm.

38 This landmark report on the potential for tree restoration to mitigate climate change was completed by the UN's FAO and the lab of Thomas Crowther. While tree planting should not be seen as an alternative to cutting fossil fuel use, the report suggested that there are 1.7 billion hectares of treeless land on which 1.2 trillion native tree saplings could be encouraged to grow. See https://science.sciencemag.org/content/365/6448/76.

39 The UN's Population Division is the authority on global population data. In 2019, it published its latest World Population Prospects, with various projections of the world population to 2100 when considering different assumptions; see https://population.un.org/wpp/. For a more digestible presentation of this data, see https://ourworldindata.org/future-population-growth.

40 For a greater explanation of Earth Overshoot Day and how it is calculated, visit https://www.overshootday.org.

41 Our World in Data is a great resource for many things, including population data. It has presentations on the growth of world population, future population forecasts, fertility rate, life expectancy and many other aspects of demography. See, for example, https://ourworldindata.org/world-population-growth.

42 Hans Rosling was a remarkable communicator of social science. His work lives on in the Gapminder Foundation; see https://www.gapminder.org/, which is full of interactive tools and videos on population and the realities of poverty.

43 For a presentation that compares China's one-child policy with Taiwan's own fall in fertility, see https://ourworldindata.org/fertility-rate#coercive-policy-interventions.

44 Both the UN Women (https://www.unwomen.org/en) and UN Population Fund (https://www.unfpa.org/) sites give thoughtful commentaries on many of these issues.

45 A detailed description of the methodology of the Wittgenstein Centre's study can be found online at https://iiasa.ac.at/web/home/research/researchPrograms/WorldPopulation/Projections_2014.html.

46 The Ellen MacArthur Foundation aims to provoke discussion and action with the ambition of bringing about a practical circular economy. Its website is a rich source of information and ideas on the subject; see https://www.ellenmacarthurfoundation.org. In addition, Kate Raworth's *Doughnut Economics* book is an insightful read as to how such a system may come about.

47 The UN's FAO 2019 report, *The State of Food and Agriculture*, included an expansive study of food waste in the world today and a review of how to reduce it; see http://www.fao.org/state-of-food-agriculture/2019. A new report, WWF-WRAP (2020), *Halving Food Loss and Waste in the EU by 2030: The Major Steps Needed to Accelerate Progress*, gives concrete guidance on how to reduce waste and is available at: https://wwfeu.awsassets.panda.org/downloads/wwf_wrap_halvingfoodlossandwasteintheeu_june2020__2_.pdf.

48 The Kigali Accord of the Montreal Protocol, signed in 2016 by 170 nations, commits governments to the correct management and treatment of HFC refrigerants at the end of life. Project Drawdown recognises this as the number 1 of the 80 climate solutions listed in their review. They estimate that it would prevent almost 90 gigatonnes of carbon dioxide equivalents from entering the atmosphere.

Index